ACHIEVING LEAN
CHANGEOVER
PUTTING SMED TO WORK

ACHIEVING LEAN
CHANGEOVER
PUTTING SMED TO WORK

John R. Henry

CRC Press
Taylor & Francis Group
Boca Raton   London   New York

CRC Press is an imprint of the
Taylor & Francis Group, an **informa** business

CRC Press
Taylor & Francis Group
6000 Broken Sound Parkway NW, Suite 300
Boca Raton, FL 33487-2742

© 2013 by John R. Henry
CRC Press is an imprint of Taylor & Francis Group, an Informa business

No claim to original U.S. Government works

Printed in the United States of America on acid-free paper
Version Date: 20121011

International Standard Book Number: 978-1-4665-0174-4 (Paperback)

Library of Congress Cataloging-in-Publication Data

Henry, John R.
    Achieving lean changeover : putting SMED to work / John R. Henry.
        p. cm.
    Includes bibliographical references and index.
    ISBN 978-1-4665-0174-4 (hardcover : alk. paper)
    1. Production management. 2. Manufacturing processes. 3. Assembly-line methods. I. Title.

TS155.H38 2013
658.5'33--dc23   2012030361

Visit the Taylor & Francis Web site at
http://www.taylorandfrancis.com

and the CRC Press Web site at
http://www.crcpress.com

Contents

Preface

I have been working in manufacturing since 1976, first in pharmaceutical manufacturing and packaging, and then selling automated packaging, assembly, and manufacturing equipment. I always realized that changeover time was lost time, but I didn't worry much about it because I did not think it was that significant and because there was not much that could be done about it anyway.

In 1993 I had dinner with a friend of mine, John Feik of DPT Laboratories, who had just bought a manufacturing company that was on life support. He spent the entire dinner telling me about his vision. He saw the path to success as being the most flexible company in the industry. He told me about Shingo's book, *A Revolution in Manufacturing: The SMED System* (1985) and how the ability to manufacture many products with minimal lost changeover time would be the key to flexibility. His company has been fabulously successful and is now the industry leader.

That was my first introduction to Shigeo Shingo, single-minute exchange of die (SMED), and the idea that changeover was addressable. The next day I ordered Shingo's book. After I read it, I ordered everything else I could find. It seemed amazing that there was so little available.

A few years later I needed a research topic for my Certified Packaging Professional (CPP) certification. I chose changeover and started digging more deeply. This paper and its publication led to a request for assistance from a client. That project led to others and thus Changeover.com was born.

At first I focused on automated packaging and assembly because those were my principle areas of expertise. In the course of my work I developed the ESEE (Eliminate, Simplify, Externalize, Execute) system. In the fifteen years that I have been helping companies with changeover, I have found that the concepts, and even many of the specific implementations, are universal across all industries. Where the implementations are not directly usable, they can often be adapted or lead to other changeover reduction ideas.

This book has three major purposes:

■ The first is to impress upon the reader the importance of reducing changeover time. Most people think of it as a tactical issue. Reducing costs by reducing downtime is always a good thing and should be done for its own benefits. Many do not realize how expensive those costs are, and this book, in Chapter 2, discusses their source and magnitude. Less widely recognized is the strategic impact of changeover. Improved changeover will improve quality. Better quality, in the form of less variability, will give any company a leg up over their competition. Even more importantly, improved, faster changeover will allow a company to be more flexible. The company will be able to produce a wider variety of products capturing market niches that may not be available to the competition. They will be able to turn these products around more quickly and make them more time responsive to their customers. This flexibility or nimbleness is something that many of the competitors, especially larger ones, are often lacking.

■ The second thing this book addresses is the cost of changeover. This is a thorny issue, and there is not a lot of information about it. It is not easy to calculate. It is not easy even to get agreement on how to calculate it. Most companies, in my experience, do not even try. In those companies that do calculate them, the costs of changeover are enormous. This book discusses some of the costs of changeover and gives a starting point for calculating them.

■ Finally, in years of collecting books on changeover, I have found that most of them focus on heavier industries. Shingo's work with SMED began with heavy stamping in Toyota and much of his work was with heavier manufacturing industries. His books reflect this. The principles discussed in these books apply to virtually all types of manufacturing and even many nonmanufacturing processes. It can sometimes be hard to see how to apply the principles to industries such as electrical/electronic, household goods, foods, pharmaceutical, and other "lighter" industries. This book is, in a sense, a translation of Shingo's principles to other industries and processes.

The ideas and examples discussed in this book have all been used successfully in a variety of applications. That they work in some applications does not mean that they will work in all applications. Use this book as a guideline, take the ideas and examples that will work in your applications, modify them as necessary, and implement them. The key is that it is a guideline—it is not an absolute blueprint.

I am always interested in hearing what has worked for you, whether it comes from the book or from elsewhere. I am also happy to answer any questions about changeover or manufacturing. Email me with questions, suggestions, or comments at johnhenry@changeover.com.

John R Henry CPP
http://www.changeover.com

Acknowledgments

Over the years, the author has received assistance from too many people to mention here. These have included client personnel ranging from top management to engineers, to mechanics and technicians, operators, and others. Many of the ideas in this book come from them. I would like to take this opportunity to thank them. It is the people on the plant floor who are the experts. I learn a lot from them. I suggest that everyone pay attention to what they have to say.

I have received assistance and ideas from machine builders in many industries. These are the people who make the production machinery. They too have come up with many ingenious ideas to make changeover simpler, more repeatable, and faster.

So to all of you, wherever you are, thank you.

Several companies have contributed specific materials for this book. They are credited in the book, but I would like to give special thanks for their support over the years. These companies and their URLs are below:

Chase-Logeman Corporation: http://www.chaselogeman.com
De-Sta-Co Corporation: http://www.destaco.com
J. W. Winco Inc.: http://www.jwwinco.com
Krones Inc.: http://www.kronesusa.com
Nail Creek Services Inc.: http://www.custommanuals.com
Oden Corporation: http://www.odencorp.com
Septimatech Group Inc.: http://www.septimatech.com
Service Engineering Inc.: http://www.serviceengineering.com
SIKO Products Inc.: http://www.sikoproducts.com
TAP Biosystems: http://www.tapbiosystems.com
Tooling Technology LLC: http://www.toolingtechgroup.com

About the Author

John Henry is widely known as the Changeover Wizard for his work in making changeover time disappear. His company, Changeover.com, has assisted many companies in a wide variety of packaging, assembly, processing, and other manufacturing industries since its founding in 1996.

After a nearly eight-year stint in the US Navy, John went to work for Alcon Laboratories as maintenance supervisor, eventually rising to manager of Facility Operations. He left to purchase Automation Sales, a company dedicated to the design, sale, installation, and service of automated packaging, assembly, and manufacturing machinery.

In the early 1990s, John realized that changeover was not a necessary evil and could be improved. He began helping some of his machinery clients and Changeover.com was born. Since then, Changeover.com has provided training and consulting assistance internationally to a variety of companies in a variety of industries.

John Henry has master's degrees in industrial management, interpersonal relations, and business education. John has been a classroom and online adjunct instructor at the Southern New Hampshire University Graduate School of Business since 1982. He teaches packaging technology at the Polytechnic University of Puerto Rico.

John has published many articles in the industry press and is a frequent speaker at national packaging conferences. He was previously a contributing writer for *Food & Beverage Packaging* magazine. He is currently a contributing writer for *Packaging Digest*. John's other books include *Machinery Matters* (2011) and *The Handbook of Packaging Machinery* (in press).

John may be contacted at johnhenry@changeover.com. He is always interested in discussing changeover.

Chapter 1

Introduction

This book is about changeover, principally about changeover of manufacturing, packaging, and assembly processes. The general concepts as well as many of the examples will be useful in other industries that require turnover of processes, such as airlines, hospitals, operating rooms, food service, and others. More specifically, this book is about converting changeover downtime to productive uptime. It is about the practical implementation of the single minute exchange of dies (SMED) philosophy developed by Shigeo Shingo at Toyota.

Quick changeover, sometimes called QCO, is a popular term for this practice, but that is not what this book is about. It may seem to be semantic nitpicking, but the use of the word *quick* in this context can be counterproductive. When teammates hear the term *quick changeover*, they are likely to hear that they will be asked to do the same thing that they have always done, but to work harder and faster to accomplish it more quickly. This almost guarantees that gains will be less than the maximum possible, short-lived, and annoying. While orientation and training can help counter this initial impression, it is better to avoid it by not using the word *quick* at all.

This book recommends the use of the term *lean changeover* (LCO) instead. Most readers will be familiar with lean manufacturing. Lean manufacturing is like lean meat. It is manufacturing from which all the fat in the form of nonproductive, non-value-adding waste has been removed. Lean changeover applies that same concept to changeover. Lean changeover is not about doing the same things faster, it is about eliminating the waste in changeover. Eliminating the waste will result in faster changeovers with less effort. Lean manufacturing philosophy focuses on eliminating the *Seven*

Wastes or TIM WOOD, as the acronym has it. Lean manufacturing and lean changeover both focus on eliminating wasted transportation, inventory, and materials (TIM) as well as waiting, overproduction, overprocessing, and defects (WOOD). Focusing on waste elimination rather than speed, per se, helps stimulate a more positive attitude. The end result is that changeover time is reduced, but on a sustainable rather than a temporary basis. The end result is that teammates see themselves as part of an effort of which they can be proud, rather than being pushed harder by management.

Years ago, Robert Heinlein wrote a short story called "The Man Who Was Too Lazy to Fail."* The story is about a "lazy" boy growing up in Appalachia. In his laziness, he reasoned that it was easier to go to school than work in the fields with his father. In his laziness, he decided to go to college rather than work in the coal mines. Laziness pushed him to become a navy pilot rather than other options requiring harder work. Because he was lazy, he invented an autopilot making it easier to fly the plane. Because he was lazy, he led a very successful life. (Need citation)

The tagline of the story is: "All progress is made by a lazy person looking for an easier way." The goal of lean changeover is laziness. Not laziness in the sense of leaving tasks undone, and not laziness in the sense of sloppy work. The goal is laziness in the sense of finding an easier, better way to accomplish changeover. When this is done, when changeover is made easier and better, the amount of time required will shrink as surely as day follows night.

---

* Heinlein, R. (1973). The man who was too lazy to fail. In R. Heinlein, *Time Enough for Love* (pp. 54–77). New York: The Berkely Publishing Group.

Definitions

Before we go on, it is important to define some of the terms used in this book. Many of these terms are in common use and many readers will probably already know what they mean. This knowledge can lead to miscommunication. The term *changeover* is fairly simple, but if 3 people are asked to define it, they may give 4 different answers. The actual definition is less critical than the fact that everyone agree to the same meaning. The definitions provided here and used throughout the book are, in the author's experience, the most common and appropriate ones. Some organizations may prefer different ones and this is fine. What is important is that everyone in the organization uses the same one, whatever it may be. Failure to do so will result in a failure of communication negatively impacting the ability to meet the goal.

Changeover

> *Changeover* is the total process of converting a machine, line, or process from running one product to another.

The emphasis must be on *total*. Some definitions of changeover focus on machine setup. These tasks may be classified as mechanical. These tasks are clearly a key component of changeover, but another group of tasks must also be accomplished in order to run the next product. These can include such tasks as closing out the production documents, reconciling raw materials with finished product, removing leftover materials from the area, picking the materials for the next product from the warehouse, line inspection, preparation of next-run documentation, and more. These tasks may be classified as operational and mechanical.

Operational tasks are every bit as much a part of changeover as mechanical tasks. They can offer as great or greater opportunity for reducing changeover time. If the mechanical portion of the changeover is performed in a timely manner, only to have the line wait an hour for materials to arrive from the warehouse, changeover is taking much longer than necessary. In this case, the best opportunity for improvement may be better scheduling of the material movement rather than improvements to the mechanical changes.

The 3 Ups

Changeover may be divided into 3 major components, called *the 3 Ups.*

Cleanup: Cleanup is the removal of all materials from the previous lot. In some cases, cleanup may be simple and consist of removing any remaining parts or materials from machine hoppers, sweeping floors, and general tidying up. In other cases, it may require cleaning or washing down of the machinery and area. In some extreme cases, such as pharmaceuticals and some foods, it may require total disassembly of the machinery, washing, sterilization, and testing.

Setup: The word *setup* is sometimes used interchangeably with the word *changeover,* but this is a mistake. Setup is a key part of changeover, but only a part. Setup consists of the physical conversion of the machinery to run the next product. In some cases, this conversion is done by adjustment. In other cases, it is done by replacing product-specific parts of the machinery, such as dies. Most machinery requires a combination of both. A capping machine on a packaging line may require adjustment of its height to accommodate a taller bottle. It may also require the replacement of the cap chute and escapement to run a different cap. These product-specific, replaceable parts are called *changeparts* in some industries, and that term will be used throughout this book. A stamping machine will require replacement of molds or dies for the next product. Its uncoiler may require adjustment to handle a different width or gauge of metal coil.

There are advantages to product-specific changeparts over adjustment, but there are also advantages to adjustment over changeparts. The pros and cons of each will be discussed in a later section of this book.

Setup will also include a number of operational tasks such as preparation of documentation, material movement from warehouse to production, quality inspection, and other tasks not directly related to the production machinery.

Startup: Startup is sometimes called *run-up* or *ramp-up.* Whatever it is called, it is that period of time after cleanup, setup, and all other changeover tasks have been performed and the line begins producing, but before it has settled down into normal operation. Startup is characterized by frequent jams, stoppages for adjustment, defective or marginal product, and anything else that prevents it from reaching normal speed and efficiency. It does not include initial line charging. In some cases it is necessary to run a line slowly, perhaps with manual operator assistance, until it is filled and all machines have a backlog accumulation on which to operate. This process, which may be called *charging* or *acceleration*, should be considered part of setup. Startup time begins when the first product for sale is discharged from the end of the line. It ends when the line is running at normal speed (averaged over time) and efficiency.

Cleanup and setup can and must be improved, but their elimination is seldom a realistic goal. The goal with startup must be elimination. It will seldom be possible to eliminate it completely because there are usually too many variables. That should not stop it from always being the goal. That should not stop anyone from doing continuous improvement to reduce it. No one should ever be satisfied with "good enough."

Startup occurs because of variation. This may be due to variations in materials or product. More frequently it is due to variation in the setup and occasionally variation in cleanup. Chapter 8 of this book will focus on driving variability out of setup. Driving variation out of setup will also have an indirect but positive impact on material variability.

When setup has excessive variation and is not under proper control, it may be hard to determine whether the variation causing startup is from variability in setup or variability in materials. Bringing the setup under control will spotlight the impact of material variability and pressure the materials and manufacturing departments to reduce it.

Variability may also be the result of material handling. In one case a process required depositing a pattern of conductive paint on a glass panel. The plant had a great deal of difficulty maintaining the consistency of the line width and initially attributed it to improper setup of the deposition system. After that system's setup was brought into control, the problem, though diminished, continued. Further investigation found that the paint was stored in a non-climate-controlled warehouse that was warm in summer, cool in winter, and seldom the same ambient temperature as the production area. The process was changed so that that the drum was brought from the warehouse in advance and allowed to stabilize in a temperature-controlled cabinet at the point of use. This resulted in a consistent viscosity and consistent application.

Paper products and components such as corrugated cases, cartons, and labels are very hygroscopic. If they are stored in a non-controlled area, they will pick up ambient moisture or may dry out. Either one can cause variability in how they run in automated machinery. Better storage conditions can reduce the amount of variability.

The purchasing department is, quite properly, under pressure to purchase the least expensive suitable materials available. The emphasis must always be on *suitable*, but what this means is not always obvious. Normally the materials will fall between an upper and lower specification for a variety of parameters. Within these limits, there is variability. Purchasing better quality, that is, less-variable materials, will usually cost more. If setup is under control, it becomes possible to determine how much startup time is caused by this variability, assign a cost to it, and determine whether it is cost justified to purchase less-variable materials.

Example: Assume that changeover time could be reduced by 10 minutes daily by purchasing a higher grade of materials. If the cost of changeover time is $10,000/hour ($\approx$$1650/minute) the savings will be:

(10 minutes × 240 days/yr × $1,650 minutes) = $3,960,000/year.

If the additional cost of higher-quality materials is $1,000,000/year, the simple payback return will be about 2.5 months. This should make it fairly easy to justify the additional expense of higher-quality materials.

The goal must always be vertical startup. The term *vertical startup* comes from graphical representations of line performance. Ideally, this graph should show zero when the line is stopped. On startup, it should shoot up vertically to the normal performance line. In reality, the line will almost always have some slope to it. The goal is to increase the steepness of that slope toward the vertical.

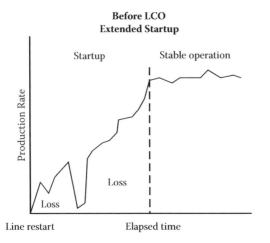

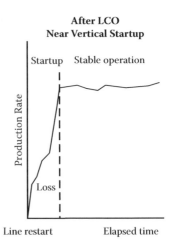

Changeover Time

When the definition of changeover is understood, the definition of changeover time is straightforward:

> *Changeover time* is the total elapsed time from the last unit of good production at normal speed and efficiency of the preceding run to the first unit of good production of the succeeding run at normal speed and efficiency.

That is a bit of a mouthful and it can be simplified to: *Changeover time* is the total elapsed time between good production and good production (with the caveat that everyone must understand that *good production* means normal speed and efficiency).

Note that it is about elapsed time, not labor hours. Even for highly skilled technicians, the fully burdened cost will probably not exceed $30 to $35 per labor hour and in many cases may be less. Changeover improvement will usually reduce the total labor hours in changeover, providing savings. Even at $30 to $35, labor hours are still relatively inexpensive compared to the cost of line downtime. The cost of an idle line will vary from industry to industry and plant to plant, but it is typically thousands of dollars and sometimes tens of thousands of dollars per hour. Lean changeover may reduce the total labor hours involved, but in some cases may not. Occasionally, it may even increase them. If an equipment washroom is the bottleneck, it may be desirable to have a person there full time. If this can reduce the elapsed changeover time, it is likely that this can be justified.

It is production downtime that is expensive—not labor hours.

Downtime

There are a number of different classifications of downtime that are commonly used in manufacturing. For simplicity, this book will use the following definition:

> *Downtime* is any time that the line should be running but is not.

Some plants will stop the line during the meal break. The line is stopped or "down," but for the purposes of this book it will not be considered downtime. Other plants will normally keep the line running during meal breaks. If one day there are not enough operators to keep running during the break due to absences, the time that the line is stopped will be considered downtime.

In most plants, the major causes of downtime are:

- Changeover
- Breakdown
- Maintenance
- Lack of qualified personnel
- Lack of quality product and/or components

These are listed in roughly the order of importance in lost hours, although this order may vary from plant to plant. The focus of lean changeover and this book is on reducing downtime due to changeover. In addition to reducing changeover downtime, lean changeover will help reduce the other causes of downtime as well.

Machines that are properly set will run more smoothly and will be less prone to breakdowns. Changeovers will be performed under less time pressure, which helps by encouraging permanent repairs to be made rather than jury-rigged temporary repairs. (Few things are as permanent as a temporary repair.) This sets up a virtuous cycle. As proper repairs are made, fewer breakdowns occur, allowing for better and more permanent repairs to be made as they do occur.

Planned maintenance, as opposed to repairs, is not as big a cause of downtime as it should be. Planned maintenance avoids unplanned repairs, which always seem to occur at the most inconvenient times. Many plants are under pressure to get production out the door and may not have time to stop the line for maintenance. LCO, by reducing the time pressure, provides more opportunities for this planned maintenance to occur.

Most plants need employees who are more than the proverbial "warm body." Training, even for low and unskilled positions, is a must. This training of personnel may take a back seat to maintaining the production flow, creating a chicken and egg situation. Training is not adequately done because of time constraints. Time constraints occur because of inefficient operation and breakdowns aggravated by lack of training. This cycle must be broken. Better changeover will free up time for more and better training. Better trained teammates will be better able to keep production flowing, freeing up more time for training. It may be difficult at first to find the time for training, but it must be done. Once it begins, it will become easier and easier to find more time for better training.

The effect of changeover on material quality was discussed previously. Lean changeover not only provides a way of justifying better materials, it also focuses on getting them where they are needed in a timely manner.

Lean Changeover is like Duct Tape
There is Nothing it can't Fix

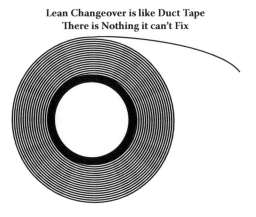

Quality

The traditional definition of quality in a manufacturing environment is usually along the lines of "meets specifications" or "within tolerance." That is, an upper and a lower specification limit for a key characteristic such as weight is established. If the product weight falls between the upper and lower limit, it is considered a quality product and released for further processing or sale. If it falls outside of the weight limits, it is rejected and discarded, reworked, or sold at a discount. In all cases, it represents a loss.

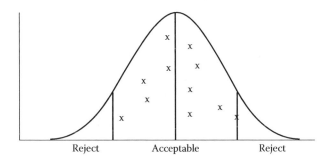

The problem with this traditional definition is that there is still variability of product between the high and low end of the specification. Consider a bottle cap that is 38mm +/–0.1mm. Assume that the bottle neck will always be exactly the same outside diameter. When applied, caps that are at the low end of the specification will be tight and hard for the consumer to remove.

Caps at the high end of the specification will be loose and may leak. If all the caps are at the low end, it is possible to adjust the torque to compensate. It is impossible to do this if they vary.

A better definition, and the one that will be used throughout this book, is:

> *Quality* is the absence of variation.

This definition at first glance appears to be different from the within-specification version. On closer examination, it is actually quite compatible. There will never be a total absence of variation. All industrial processes will always vary. The traditional definition implies that once in specification, no further improvement need be made. The absence of variation definition insists that continuous improvements be made to bring the process or materials ever closer to that point of zero variation. This is really what the customer wants, whether the customer is a downstream process or the end user. They want it to be exactly the same every time. As a Holiday Inn ad campaign put it years ago: "The best surprise is no surprise."

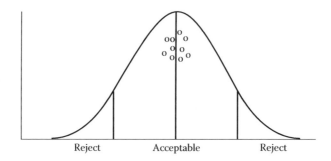

An adjunct to this definition is the Taguchi Loss Function. Genichi Taguchi developed this model to show that any time there was variation, there was an associated cost that increased the further away from perfection the process was. This cost of absence of quality could be calculated. This allowed cost–benefit analysis of non-perfection.

Importance of Changeover

So why is changeover so important now?

Fifty years ago Coca-Cola was available in a single flavor, now called Coke Classic. It was available in a single format: a 6-ounce glass bottle. The bottling plant manager had an easy life, at least when it came to changeover. They could set up the bottling lines and retire 40 years later never having had to perform a changeover.

This may be an extreme example, but it was closer to being the rule than the exception. Many products were manufactured in limited varieties that changed only slowly over time. Long-lived products meant that plants could become expert at running them. Limited variety allowed for long production runs with few changeovers. When changeovers did occur, their duration didn't much matter in the grand scheme of things. Competition, especially international competition, was limited, which reduced incentives to cut costs. Inefficiencies could largely be passed along to the final customer. Since all companies had similar inefficiencies, customers had little alternative but to accept it.

This has changed radically in the past 30 years or so. Consumers are demanding more variety. Where there was a single Coca-Cola 50 years ago, there are more than 150 varieties of Coca-Cola today. It is available in at least 6 different flavors (i.e., Classic, Diet, Caffeine Free, Lemon, etc.), each of them available in cans, glass, and plastic bottles of various sizes. In addition to the varieties of Coca-Cola, the typical bottler will also produce a variety of other products, including Sprite, Fanta, lemonade, and more. This has taken the bottling plant from never doing changeover to performing multiple changeovers per shift in some cases.

Oreo cookies used to be available in a single size in a limited range of packaging. They are now available in several dozen (at least) varieties. Tylenol is available in over 100 varieties and packages. A multinational battery manufacturer runs over 750 stock-keeping units (SKUs) in one of its plants.

All of this contributes to frequent changeover. Changeovers must become a core competency in all plants.

This diversity of products is going to get worse in coming years, not better—worse for manufacturers, that is. Consumers are no longer satisfied with *one size fits all*. They never really were, and it never really does, but now the consumer has choices. One company may make their product available in 4- and 12-packs. One of these may be suitable for many consumers, but there will be some for which 4 is not enough and 12 is too many. This may cause those customers to jump from this product to a competitor that provides an 8-pack. Consumers are forcing manufacturers to move toward mass customization. Mass customization is the mass production of customized products to fill market niches. Lot sizes are being reduced from hundreds of thousands of units to thousands, hundreds, and in some cases to a single unit. It is not enough to produce in low volumes;

costs must be slashed to allow those low volumes to be produced as efficiently as the former high volumes.

Another factor related to consumer preferences is changing distribution. There used to be a few channels for any product. Breakfast cereal used to be sold primarily through traditional grocery stores. Now it is sold in grocery stores, convenience stores and gas stations, pharmacies, wholesale clubs such as Sam's Club and Costco, dollar stores, and more. Each will have their specific requirements as to size and presentation. Instead of the cereal being sold in 2–3 different packages, a dozen may now be required.

The days of long production runs with a changeover every week or two are long gone and never coming back. Many plants do changeovers daily and some find themselves doing multiple changeovers per shift. With fewer products over which to amortize the cost, as well as less excess production capacity, changeovers must be performed as quickly and precisely as possible.

Pit Stops

One useful model for changeover is the NASCAR pit stop. Some people find this to be the most interesting part of the race. In the typical pit stop, the car will have 4 tires replaced, fuel added, the windshield cleaned, perhaps some adjustment to the motor or suspension, and drink given to the driver. NASCAR does this in about 12–13 seconds. Formula 1, with different rules and techniques, does it in 7–10 seconds. Pit stops make a good analogy for manufacturing changeover as both are very similar in a number of ways.

The first similarity is the motivation for reducing pit stop and changeover times. The motivating factor in both is money. When the car is not on the track, it is not competing. In most cases, cars are pretty evenly matched and it is not uncommon to see less than 50 feet separating the 1st and 5th place cars at the end of a 400 mile race. As all cars must make pit stops, one way to gain a competitive advantage is to spend less time at it than the competition. At a speed of 150 miles per hour, 1 second is equal to 220 feet of track position. Even seemingly insignificant reductions in pit time can give enormous advantage on the track. This translates to enriched paybacks in prize and sponsorship money.

A manufacturing plant is no different. When a line is down for changeover or any other reason, no product is being produced. More importantly, no money is being made. Reducing the time the line is down for changeover will increase the time the line is making money. The costs of changeover downtime will be discussed in detail in Chapter 2.

As in the pit stop, small savings in changeover time can produce big results. Some might ignore some inefficiencies as being too small to matter. In reality, nothing is too small to matter. To put it in perspective, a savings of only 10 minutes per day in changeover will, at the end of the year, result in a full week (40 hours) of additional production. In most cases, this additional week of production costs little or nothing. It is almost pure profit.

A similar effect holds for seemingly small increases in production efficiency. A process running 100 parts per minute (PPM) at 80% efficiency will produce 9,600,000 parts in a 2,000-hour year. If efficiency can be increased to 85%, the same process will produce an additional 600,000 parts at no additional cost other than materials.

The reasons for improving both pit stop and changeover are the same. So are the techniques for achieving it.

- *Task assignment:* Pit crew members know exactly what their assigned tasks are and the order in which they are to be performed. There is no task left undone nor is there any double effort.
- *Training:* Pit crew members know exactly how to best perform each task.
- *Specialized tools:* Pit teams use specialized tools, such as super-high-speed impact wrenches, to allow them to perform more efficiently.
- *No tools:* No tools are even better than specialized tools. Tools can cause all sorts of issues, from the wrong tool being used, to the tool not being found or being damaged. One example is the use of peelable film to eliminate the need for a sponge and squeegee to clean the windshield.

- *Prepreparation:* When the car enters the pits, all materials, tools, and parts have been staged so no time is lost fetching them.
- *Monitoring and tracking:* Each and every pit stop is monitored and the times tracked. Slow-motion cameras allow every wasted motion to be identified and eliminated.
- *Continuous improvement:* Pit crews work to continuously improve the speed and quality of the pit stops. They do this by evaluating each stop to determine what was done right and what can be done better. Mistakes are noted and means developed to avoid them in the future. The evaluation is done in full knowledge that they are unlikely to find any major improvement opportunities. In most cases the improvements will be marginal. These marginal improvements, taken together, become cumulatively significant.

All of the above and more apply equally to changeovers and will be discussed in more detail in succeeding chapters.

History of Changeover

Changeover has been recognized as an issue for more than 100 years. Frederick Taylor touched on it in his 1911 book *Shop Management,* when he discussed unproductive time in setting work (as he called it) as well as the need for more precise and repeatable setting. Henry Ford, in his 1923 autobiography, recognized that changeover was unproductive. His policy was never to do changeovers. If there were two parts, he would install two machines.

Until the 1960s that was pretty much the state of the art. Changeover was best avoided by long production runs. Limiting the variety of products made was an important part of this strategy. Some nibbling about the edges might be possible, but changeover was viewed as basically a necessary evil that had to be accepted.

This changed in the 1960s at Toyota. They were making a limited range of cars and felt that if they could make a wider variety of models they could break out of their national market into the international marketplace. The problem with increasing variety was changeover, more specifically the setup of the stamping presses that made body panels such as doors and fenders. At the time, changing the dies from one model to another took 10–12 hours.

Shigeo Shingo, a Japanese industrial engineer, was contracted to analyze the problem and improve die changeover. He had been working at reducing changeover and setup times at other companies, but it was at Toyota that he perfected his ideas. It was his recognition of two key factors that formed the basis for what he called the SMED system.

First, much of the work involved in changing dies was done with the press stopped. Shingo recognized that press time, compared to labor time, was expensive. If some of the tasks could be performed while the press was running, it would cut the changeover downtime. Shingo began by using standard industrial engineering techniques to break die change into its constituent tasks. It is difficult to evaluate a die change. It is much simpler to evaluate how a fastener is tightened. He then classified all tasks into *internal* and *external* (sometimes called *intrinsic* and *extrinsic*). Internal tasks can only be performed with the press stopped. External tasks can be performed while the press is running, either before or after the changeover. Removing the bolts mounting the currently running die is an internal task, as it can only be done after production is complete and the press is stopped. Staging the next set of dies, along with tools, fasteners, and anything else required, may be done prior to completion, while the press is running, and is an external task. Likewise, as parts were removed, they were set aside next to the press and not moved to storage until after the press was running again.

Shingo's second great insight was that much time was lost setting the press to accommodate different die sizes. He realized that if all dies were the same size, this adjustment could be eliminated. In this case it was prohibitive to make all dies the same size so he did the next best thing. He permanently mounted all of the dies on baseplates. These baseplates were all the same external size at the mounting points and all used the same fasteners in the same locations. The thickness of the baseplates was varied so that the combined dies and baseplates all had the same overall and shut heights. This eliminated the need for any adjustment in mounting. It also eliminated the need for multiple sizes and types of mounting fasteners, which Shingo had also identified as a time waster.

Standardization of die sizes had a cascading effect and brought many benefits beyond the obvious ones mentioned previously. Eliminating the need for adjustment and setting meant that there would be much less variation between setups. Setup would always be done correctly, which meant that the first piece off the press would be correct, reducing waste. Reducing variability between production runs meant that there would be an overall better fit between parts when they were assembled. Simplifying and standardizing the setup allowed machine operators to perform changeover, which improved their skill sets and made them more effective machine operators. Standardizing the setup meant that highly skilled mechanics were no longer required for routine die exchange and setting. This freed them to perform more valuable work throughout the plant. More importantly, it freed them to develop and implement further improvements.

The end result was that Shingo was able to reduce press changeover times from 10–12 hours to less than 10 minutes. As described here it seems rather simple and obvious. In reality, Shingo's discoveries took many years of hard work to develop and implement.

Shingo called this system SMED for single minute exchange of dies. At first glance the title may seem misleading. Many might think that *single minute* means one minute or less. A worthy goal certainly, but not what Shingo had in mind. By single minute, he meant single digits or less than 10 minutes. He shared his insights and methodologies in an excellent book, *A Revolution in Manufacturing: The SMED System.* This is a must-read for anyone interested in changeover. A companion volume, *Quick Changeover for Operators,* covers much of the same ground but in a greatly simplified format. It is aimed at the shop floor operators, mechanics, technicians, and others. Other works by Shingo are available from Productivity Press and are equally worth reading.

SMED gave Toyota more flexibility and capacity, freeing them from the need for long production runs. It is not an exaggeration to say that much of the Toyota Production System, such as just-in-time, Kanban, continuous improvement, 5S, and autonomous maintenance was facilitated by Shingo's work in changeover.

Tactic or Strategy?

Lean changeover is certainly a good business tactic. It will reduce costs in much the same way as turning off unnecessary lights will. The cost of changeover downtime, discussed in depth in Chapter 2, is usually thousands or tens of thousands of dollars per hour, and any reduction will have significant impact on plant operating costs.

More importantly, lean changeover is a strategy that can and has allowed companies to leapfrog their competition. Toyota is one famous example. Another company was a failing division of a large multinational scheduled for shutdown in the late 1980s. The plant manager bought the division and established it as a standalone company. His vision was to make it the most flexible company in the industry. SMED and the ability to run a large number of products in small production runs were a key part of that vision. Today, that company is number one in its large industry segment. Another nationally known company produced over 800 SKUs in one of its plants. Their marketing department decided that they would produce these to order rather than to inventory. It would also offer 48-hour order-to-shipment turnaround. This necessitated as many as 4–6 changeovers per shift on each of 5 production lines and caused a great deal of production pain in the plant. The results made it worthwhile. The 48-hour strategy was credited with an 8% increase in national market share. They eventually learned how to do lean and fast changeovers and now do them almost on the fly.

Improved manufacturing flexibility and reduced costs allow a manufacturer to better satisfy their customers in three important dimensions: Product variety, responsiveness, and price. It also allows manufacturers to better satisfy their stockholders and stakeholders through improved profitability and increased shareholder value.

Lean changeover will reduce costs while allowing greater volume. Reduced costs allow lower prices, which fuel sales demand. As volume increases, fixed costs are amortized over more and more units. This allows still lower prices, fueling further demand. As the shampoo label says: "Rinse and repeat." Many have heard of the vicious cycle; this is its nicer cousin, the virtuous cycle.

The ESEE Concept

A clear roadmap is a requirement for any productivity improvement program, and lean changeover no exception. When you don't know where you are going, you will probably wind up somewhere else. This book will use the ESEE (pronounced easy) model for changeover developed at Changeover.com. ESEE identifies the four key areas of lean changeover and is an acronym for Eliminate, Simplify, Externalize, Execute. The words are fairly self-explanatory:

Eliminate any unnecessary or non-value-adding tasks.
Simplify everything as much as possible.
Externalize any tasks that can be done during production.
Execute by performing all changeover tasks repeatedly with minimal variation.

Each of these steps will be discussed in detail in subsequent chapters.

Today's marketplace is continually being chopped into smaller and smaller segments. Each segment demands that products be tailored to their particular wants. This is great for consumers but puts a large burden on manufacturers. The company that produced 10 variations of their product 20 years ago may be producing 40 variations today. Twenty years hence they may be producing 100 variations.

The customer is not only demanding custom tailored product, they are demanding that it be supplied more cheaply and more quickly than before. Failure to do so will result in losing that customer.

To survive and thrive, manufacturers must be able to meet these challenges. Lean changeover is an essential tool. This book will show in theory, as well as practical example, how to achieve it.

Chapter 2

Changeover Costs

Knowing Your Costs Is Critical

Many plants do not seem to know their costs of downtime or changeover time. This lack of knowledge makes it hard to justify the necessary expenses of reducing changeover or evaluate the results after. This forces justifications to be made in terms of "an investment of X dollars provides a return of Y minutes." Absent a conversion factor for minutes to dollars, it is very hard to convince management to spend money.

A key first step in any changeover program is identifying the costs of changeover downtime. In general, they will be similar to the costs of any other downtime, with perhaps a few differences such as labor costs and costs of product used during setup.

Example: A project proposes to reduce changeover time by 10 minutes per day by purchasing an additional set of changeparts at a cost of $50,000. This will allow externalization of parts cleaning. Management is likely to look at this, think "It's only 10 minutes," and disallow the expenditure. They would probably be quite right to do so on the evidence presented; 10 minutes does not seem very significant. If the cost of changeover downtime has been established to management's satisfaction as $12,000 per hour, this means that the additional set of parts will produce a savings of $2,000 per day or $500,000 per year (based on 250 days per year). Management will likely be thrilled at the one-month payback.

The dollar cost must come from the finance or accounting departments. Engineers, production managers, operations specialists, and others can certainly make the calculations and they will generally be correct. This does not prevent them from being viewed with suspicion. Unless the costs are calculated by, or at least reviewed and approved by finance, they will never be viewed as official and it may be difficult to get them approved in justifications.

If it is not possible to get a dollar figure from accounting, an alternative is to express the time savings cumulatively. If an improvement will eliminate 10 minutes daily, this does not look like much. Express it in annual terms and it becomes quite impressive. Ten lost minutes per day is more than 1 shiftweek of lost production per year (10 minutes × 250 days/60 minutes/hour = 41.6 hours). Is gaining an additional week of production capacity worth $50,000? This is an easier decision for management to make than deciding whether 10 minutes is worth $50,000.

Changeover costs take two forms, tangible and intangible. Tangible costs are those that are relatively easy to measure quantitatively. The cost of labor involved in changeover is a tangible cost and is the product of the number of teammates involved times their hourly costs. Intangible costs are costs that are difficult or impossible to assign a dollar value. That does not mean they are less important than tangible costs. In some cases they may be even more important in terms of amount and impact on the business. *Intangible* only means that they are more difficult to quantify.

Any discussion of costs in a book such as this will be of necessity generic. Not all factors will apply to all plants. A plant that is running at 50% capacity utilization will see little benefit from increased capacity. Their metrics might actually show a decline since the utilization would be on a larger available base capacity. Other plants might have additional costs not covered here. This discussion is presented for illustration of some of the more common costs of changeover.

WIIFM

A potential issue with lean changeover (LCO) programs is that the teammates, especially those on the floor, may see it as an additional imposition on them. They can see the benefit to the company in the form of reduced costs and increased profit. It is harder to see how it benefits them. LCO project leaders always need to look at each team member and imagine the letters WIIFM engraved on their foreheads. WIIFM stands for "What's in it for me?" There are direct benefits to the individual teammates. They are fairly clear once explained, but they may not be obvious without explanation. Never assume that they will figure it out by themselves. Some will, some won't. Understanding the benefit will go a long way toward getting them to accept the program.

One of the goals of LCO must be to involve operators more intimately in changeover. At first glance, they may see this as loading more work on them in addition to their normal job. One of the benefits to LCO is that changeovers are done in a more organized and precise manner. Some rudimentary maintenance may be done by the operators during changeover, such as cleaning, lubricating, and inspection. This will improve machine operation. Machines will run more smoothly and the operators will not need to spend the shift fighting the machines. This makes the operator's day easier and less stressful. Smoother-running machines mean that the mechanics will spend less time fighting fires. This allows more time for them to work on more value-adding tasks.

Any time an operator or technician must intervene with a machine, to adjust it or clear a jam, there is a possibility for an accident to occur. This must never be the case. Safety must *never* be compromised. Unfortunately, in the real world of the plant floor, it occasionally is. This can be compounded in plants with high turnover in which teammates may not get adequate training or experience in working safely. Improved changeover, especially improvements in the precision of the changeover, will reduce the number of incidents requiring intervention with the machinery. Reducing intervention will improve safety by reducing the opportunity for accidents to occur. Note that this is no substitute for adequate safety procedures, training, and enforcement. Safety must always be the number one goal. Still, nothing is perfect and reducing opportunities for accidents will help reduce them.

Some plants use mechanics to perform equipment changeover. Time spent on changeover is time not spent on preventive or repair maintenance. Time spent on changeover is time not spent on developing and implementing improvements. Time spent on routine, repetitive tasks such as changeover is time spent not fully utilizing the talents and skills of the mechanics. LCO will free up their time and allow better utilization. Most mechanics will find that this makes for a more interesting and challenging workday.

Most people like interesting challenges. Involving all teammates in LCO rather than imposing it on them can present these challenges. Teammates will find it interesting to be involved in improving their workplace. As they get more involved, they gain higher-level skills that increase their value to the company and in the marketplace generally.

Teammates are not just employees, they are partners. Their financial future is tied to the company's financial health. The better the company does, the better they will do. A profitable plant can stay in business. An unprofitable plant will eventually close. Improved plant efficiency and profitability from improved changeover will result in improved job security. Teammates often have direct participation in profits through 401Ks or other profit-sharing plans. There are never any guarantees, of course, and when presenting this, care must be taken not to inadvertently instill a sense of guarantee.

In a similar vein, a profitable plant is more likely to grow as business increases. This company growth will provide opportunities for promotions and personal growth.

There are a number of benefits to the individual from the LCO program. As part of the program, these benefits must be discovered and explained.

Tangible Costs of Changeover

Lost Production

The most obvious cost of changeover is lost production. When a line is not running, it is not making product. More importantly, it is not making money. Note that phrasing. American English is one of the few languages where the phrase *making money* is routinely used—so routinely in fact that we seldom pay attention. In reality, this is what a manufacturing process does; it creates value or literally *makes* money. Most people see products coming out of the end of the process. They need to be trained to think of them as money. They need to realize that those products go out into the marketplace and are exchanged for money. A portion of that money returns to them in the form of wages and benefits.

How much money is made will vary depending on the product and production rate as well as the value of the individual product. Some companies use lost sales. This is probably not realistic in most cases. A better value would be the contribution (sometimes called *margin* or even *profit*) of the product at that point. This book will not get into a discussion about how to measure the value or what value should be used other than to say that it must be determined by the finance department. Once the per-unit cost is determined, determining the cost of lost production is fairly simple.

Assume a per-unit contribution of $1.00 at the end of the line. If the line speed is 200 parts per minute (PPM), the lost contribution will be $1.00 × 200 for a total of $200 per minute or $12,000 per hour.

This is a bit simplistic and assumes that a plant is running at full capacity. At full capacity, no additional product can be made and any losses in production will represent a permanent loss. In many cases, lost production can be made up with overtime, additional shifts, or line speedup. A plant that is running at significantly less than full capacity may not even have enough work to fill the normal work day. In this case, lost production time from changeover has essentially no production direct cost.

Even so, it is still not totally free. If lines are being underutilized, freeing up the teammates with reduced changeover will allow them to perform other tasks such as cleaning, maintenance, training, and particularly developing and implementing improvements.

Lost Capacity

Lost capacity is the other side of the lost production coin. If a plant is running at 100% capacity, when the marketplace demands more product, the only way to supply it is by purchasing additional production equipment, hiring additional people, and possibly constructing additional manufacturing space.

Reducing Changeover Times Creates Additional Capacity

Assume a plant that is operating 24/7/365 and is completely maxed out on capacity. Line speed is 200 PPM with 30% downtime for changeover.

Total annual production is
$$200 \times 60 \times 24 \times 365 \times .70 = 73,584,000 \text{ products per year}$$

If changeover downtime can be cut in half, usually a very reasonable goal, total output increases to $60 \times 24 \times 365 \times .85 = 89,352,000$ products per year. This is an increased output of 15,768,000 products per year with little or no capital investment other than what might be required for the LCO program.

Impact on Inventory

One popular way to reduce total changeover cost is to reduce the number of changeovers by increasing lot size.

The drawback with this approach is that it results in larger finished goods inventory. Inventory is very expensive. One way to think of it is as an interest charge, typically around 30%, based on the average value of the inventory. A plant holding $5,000,000 worth of inventory will pay about $1,500,000 per year in carrying or holding costs. Reducing average inventory levels will reduce those annual inventory carrying costs.

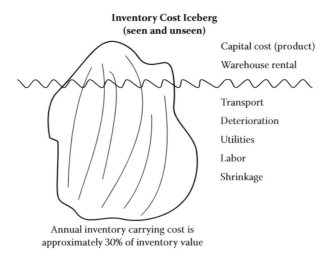

Changeover is a fixed cost. It will be the same amount whether the lot to be run is one unit or one million. Companies may attempt to reduce their changeover costs by running larger lot sizes, spreading the cost over more units, and thus reducing the changeover cost per unit. The downside to increased lot size is increased inventory size and increased inventory carrying cost. This chart shows a rather simplistic inventory model to illustrate this point.

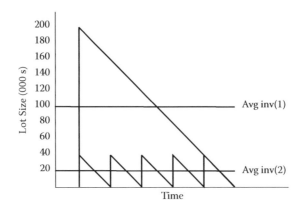

The company needs to produce 200,000 units of this product per week. It can produce a single large lot. If shipped uniformly over 5 days, it will result in an average inventory level of 100,000 units.

If the value of each unit is $2, the total annual inventory carrying cost will be $60,000.

The second option is to produce 5 production lots of 40,000 units each. Total production is the same. All units are shipped uniformly over 5 days. Average inventory is 20,000 units and annual inventory cost is $12,000. This represents an annual savings of $48,000 per year. Multiple products will multiply the savings.

Savings from reduced inventory is offset by the increased costs of more frequent changeover. Many companies will use some variation of the economic lot size (ELS) model to balance the cost of changeover and inventory costs. The ELS model shows the cost of inventory increasing with lot size and the cost of changeover decreasing with lot size. Total cost is the sum of the two. The ideal lot size is where the two costs intersect.

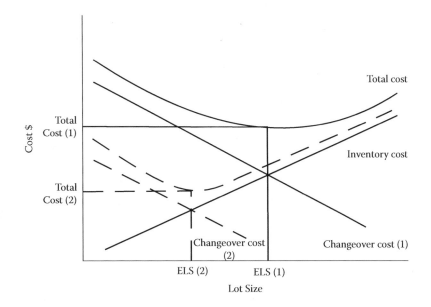

The chart shows that reducing the changeover cost will allow smaller lot sizes to be run, reducing inventory and thus the total cost.

Those more deeply involved in inventory systems may find these charts a bit simplistic, and they are. The intention is to illustrate the general shapes of the curves rather than all the possible permutations that they might take. In real life, as the saying goes, fixed costs vary and variable costs don't.

Whatever the exact shape may be in reality, lower changeover costs will generally result in lower economical lot sizes.

Another caveat is that other factors can impact optimal lot size. Mixing tank size, production capacity in an 8-hour day, regulatory requirements, and raw material packaging are a few of the things that can impact the final decision on what size lot to produce.

Reduced inventory will free up plant floor space. This space may be usable for additional production capacity or other productive uses. If not, some use should be found for it. If the space is left open, it will gradually fill with inventory again. One company, after freeing up warehouse space, elected to use it for offices for supervisors. Their thinking was that if offices were placed in the vacant area it would be difficult for it to become warehouse again. Absent warehouse space, there would be less ability for inventory to grow back.

Smaller lot sizes may also permit smaller inventories of raw materials and especially of work-in-process (WIP) inventory. The effect is less direct, but any reduction in these inventories, especially WIP, is a good thing.

Parkinson's Law states that "Work expands to fill the time available." The corollaries are that pretty much everything else expands as well. Inventory will expand to fill the warehouse space available.

Obey the Law.

Labor Costs

Labor costs of changeover may be both direct and indirect. Direct costs will mainly be the cost of the mechanics and operators who perform the changeover. In some cases, outside contractors such as a janitorial service may be used to provide part of the changeover. The number of labor hours expended carrying out each changeover is relatively easy to determine. The dollar rate for these hours is easy to determine. It is important to note that the relevant rate, in the case of employees, is not their hourly pay. The rate that must be considered is the cost of having them on the payroll. This includes their wages, but also includes the additional burden of vacation and other time off, insurance, uniforms, and other benefits. These typically increase the hourly rate by 25–50% over the stated wage rate. A mechanic earning a wage of $20/hour will have a total cost of $25 to $30/hour. This is the relevant cost that must be used in any cost–benefit calculations. The $20/hour is mainly relevant only to the teammate.

In addition to the teammates on the floor, there are usually other team-mates off the floor working on the changeover. These include clerical personnel closing out the previous production order and preparing the next one, washroom personnel engaged in washing equipment, and warehouse personnel to name a few. The time they spend working on changeover may not be, and often should not be, simultaneous with the time spent on the floor on changeover. These operations should be externalized to the maximum extent possible and should be done while the line is running. The concept of externalization will be covered more thoroughly in a subsequent chapter. Whether the work is done simultaneously with the changeover or externalized, it must still be measured and charged to the changeover. The work may be out of sight, but the costs are still out of pocket.

After measuring the hours spent by all direct labor, and establishing their rates, it is easy to determine the labor cost of changeover time: Assume that 2 mechanics at $30/hour plus 2 operators at $20/hour spend 2 hours on the floor working on the changeover. They are supported by a clerk who spends 30 minutes on documentation at $20/hour, 1 washroom person who spends 1 hour at $20/hour, a warehouse worker who spends 1 hour at $20/hour, and a quality inspector who spends 15 minutes at $20/hour approving the line for production.

Total labor cost of the changeover is as follows:

$$2 \text{ mechanics} \times 2 \text{ hours} \times \$30 = \$120$$

$$2 \text{ operators} \times 2 \text{ hours} \times \$20 = \$80$$

$$1 \text{ clerk} \times 0.5 \text{ hours} \times \$20 = \$10$$

$$1 \text{ washroom operator} \times 1 \text{ hour} \times \$20 = \$20$$

$$1 \text{ warehouse worker} \times 1 \text{ hour} \times \$20 = \$20$$

$$1 \text{ quality inspector} \times .25 \text{ hours} \times \$20 = \$5$$

$$\text{Total changeover labor cost} = \$255$$

The actual wage of each employee will often vary slightly. Individual costs can be used, but it will probably be more convenient to use average or generic amounts for classes of teammate.

Management may say that these teammates are on the payroll regardless of whether they are performing changeovers or not. This is a valid point. Their cost is relatively fixed and reducing the amount of time spent on changeover will not reduce it. The answer to this is that when they spend more time on changeover than necessary, that time is wasted. Excess time spent on changeover is time that they are not performing maintenance, making permanent instead of jury-rigged repairs, cleaning, organizing, training, and performing many other valuable tasks that are not currently performed due to lack of time.

There will also be indirect labor involved in the changeover, mainly supervision. These costs are more difficult to attribute to the changeover.

Product and Material Losses

Product and material is lost at all stages of changeover. At the end of the run, product will remain on the line that may need to be discarded as part of cleanup. During setup, it is frequently necessary to use product and materials for testing or setting. In most cases this is discarded. Once the process is restarted, there is the setup period where there will be higher-than-normal rejects. The cost of materials used in changeover can be significant. One of the focuses of lean changeover must be to reduce these losses.

Intangible Costs

Intangible costs can be hard to see and even more difficult to measure. That does not mean that they are not significant. They are, and must be recognized and addressed.

Response to the Customer

The most important intangible cost is a decreased ability to respond to customer desires. This may take the form of preventing growth of the company. It may take the form of losing market share to the competition. In some cases it may take the form of destroying the company and its jobs.

For better or worse, Walmart is a major customer of most US manufacturers. Many companies rely on them for 30–60%, and sometimes more, of their total sales. When Walmart calls and asks for speedier delivery on an order because sales are better than expected, it is virtually impossible to say no. The company that does this is likely to find that they are no longer a Walmart supplier, with disastrous consequences. Companies will work overtime, delay other customer orders, juggle schedules, and do whatever else is necessary to break into an existing production run to produce Walmart's order, but this too has negative consequences and costs. Chief among these is that satisfying Walmart may be done at the expense of dissatisfying other customers, both large and small.

Improved changeover not only avoids negative costs and lost business, it provides positive strategic benefits. The superior company will already be making small production runs with short, easy changeovers between them. Shifting from one product to another will be all in a day's business. The ability to shift between products and make small production runs protects existing markets. It also brings in new customers from less-responsive competitors. It can even open new markets that did not previously exist.

One water bottler in California started in a garage in the early 2000s. Their vision was to produce large volumes of house-brand water for supermarkets and other sales channels as well as small volumes for outlets such as gyms that wanted their own brands of water. They have developed their process so that they can run custom flavors, custom labels, and even custom bottles with minimum order sizes as small as 36 cases. At the same time, they can run the large-volume products for the supermarkets and other channels at competitive costs.

A consumer electronics manufacturer decided to offer 48-hour shipment on all of their 800+ products and was able to increase national market share by 8% in a highly competitive, almost commodity market.

Single-minute exchange of die (SMED) was developed at Toyota specifically for the purpose of better responsiveness to the market with a wider variety of styles and options. Toyota is the leading automobile manufacturer today. They are also recognized as excellent practitioners of all the lean strategies including lean changeover.

Capacity Utilization

Total manufacturing cost is strongly influenced by capacity utilization. Capacity utilization is strongly influenced by sales volume, which is strongly influenced by selling price. Selling price is strongly influenced by manufacturing costs, closing the circle. This creates a spiral that can be ridden up or down. Increase capacity utilization and total costs will decline, permitting a decrease in selling price. The decrease in selling price will increase sales volume and capacity, reducing costs further, and so on.

The alternative is to ride this spiral upward. Increased selling price reduces volume and increases costs, puts upward pressure on selling price, and further reduces volume and so on until the company dies.

This virtuous cycle was one of the great insights of Henry Ford a century ago. He realized that if he drove costs down, he would drive sales, and production, up. This permitted further cost reductions and greater sales. This tendency was so strong that many times Ford did not wait for costs to come down before reducing the price further. He relied on two effects: The first was psychological. If the cars were selling below cost, it placed tremendous pressure on the plant to get the costs down. If Ford had waited for the costs to come down before reducing prices, the plant might have come up with any number of excuses to delay developing and implementing cost saving techniques. Second, Ford realized that increased sales would create its own cost savings.

LCO, by increasing capacity, helps put a plant on the downward side of this spiral.

Quality

Quality is the absence of variation. It is not just being within specification, it is being consistently the same—product after product after product. Improving the changeovers will result in machines that are set more precisely. Reducing changeover time will allow more time for maintenance and permanent repairs. Better-tuned machines will operate more precisely and consistently.

Finally, the pressure to produce can lead to shortcuts being taken and even marginal product being accepted. This should never be the case, but it does happen in some plants. The ability to work more slowly and calmly will help prevent this temptation.

Stress on People, Machines, and Systems

When plants have excessive changeover times, it tends to stress the people, machines, and systems. Stress on people can cause job dissatisfaction leading to increased turnover and decreased performance. Stress on machines takes the form of them being worked harder and longer with less maintenance than they need. This results in more failures.

Reducing time wasted in changeovers will reduce stress levels.

Reduced Innovation

IBM ran an ad several years ago with a picture of a harried executive and the caption "Innovative thinking? We don't even have time for bad thinking!" In how many plants is this the case? Teammates at all levels instinctively know that there are many improvements that could be made to their process. While they know this, they may not have the time to think them through and develop a plan and justification. Worse, even if they can develop them, they may not have the time to implement them. Companies have to take some risks. Not all improvement ideas will be home runs. Some of them, no matter how carefully thought through, will not pan out at all. They still need to be tried. For many decades, Babe Ruth held the major league baseball record for home runs. What most people do not realize is that for many years he held the record for strikeouts as well. It would be nice if continuous improvement generated nothing but home runs. Unfortunately, it will not, but it still must be practiced on a daily basis. LCO, by freeing up resources as well as giving all teammates a better understanding of the entire process, will permit more improvement ideas to be developed and tried.

Conclusion

The importance of knowing the cost of downtime in general and changeover downtime specifically cannot be overemphasized. The costs are huge but in most plants seem to be undefined. They are not easy to calculate. They may change depending on the product, production schedule, or other factors. It is not necessary to have the costs figured down to the penny for every changeover. What is needed is an average cost per hour of changeover that can be used to justify and motivate changeover improvement.

Chapter 3

Standard Operating Procedures

An orchestra is a collection of musicians who must be closely coordinated in order to play beautiful music. Coordination is achieved via the use of a musical score. The score tells them what to play, how, and when. A conductor keeps them all properly on task. Absent the score and conductor, the likely result would be a cacophony of noise rather than beautiful music. A changeover is analogous to an orchestral piece in this regard. The standard operating procedure (SOP) is the score by which all participants know what they are to do, how, and when. The supervisor is the conductor who keeps them on task. As with an experienced orchestra, a well-trained changeover team may not need much instruction and supervision. They always need some. It is only by this means that proper and efficient changeovers can be performed.

Standard operating procedures or SOPs go by a number of different names, including standard operating instructions, procedural guidelines, or setup specifications to name a few. For simplicity, this book will refer to them as SOPs. They can be presented in a variety of formats including textual, pictorial, schematic, flow charts, or a combination. Whatever they are called and however they are presented, they all have, or should have, one goal: to properly guide the mechanic or operator through the changeover. SOPs are sometimes confused with checklists, but they are two different things. This chapter will discuss both and demonstrate why both are required.

Absent an SOP, changeover is carried out according to the judgment and experience of the teammates. Training may be done by having the new teammate observe a more experienced colleague. As this type of training passes from person to person, a great deal of clarity is lost. This will lead to the continuation of poor practices. Some will be able to perform a good changeover, others will not. In all events, the way the changeover is performed will vary from person to person. In many cases, even the same person will not perform the changeover exactly the same way twice. This leads to slow changeovers and improper changeovers. Improper changeovers will at best result in excessive startup times. At worst, they can damage machines or product and cause further downtime while the damage is corrected.

A checklist is not an SOP. A checklist is an abbreviated version of the SOP. It must be considered as a supplement to the SOP. A good way to describe the difference is that a checklist says what is to be done. The SOP explains how to do it. The goal with SOPs must be to make them detailed enough that anyone with general knowledge can work their way through it and successfully perform a routine changeover with little or no specialized training. The author realizes that in actual practice this may not always be realistic. That should not prevent it being a goal to strive for.

This level of detail can make for an SOP document that may be too long for convenient daily use. This is where the checklist comes in. The checklist serves as the abbreviated summary that the experienced person uses daily to assure that the changeover is performed correctly. In the event of uncertainty, they can refer back to the SOP for detailed instructions.

Clarity is a key distinguishing factor of a well-written SOP. Each company needs to develop a uniform format for all SOPs. A uniform format, with all information always in the same place, makes them much easier to use. It also makes it easier to pick out important information. The template discussed in the following pages is based on the style used in the pharmaceutical industry. An annotated version of the SOP appears in the appendix to this book. A blank version of it may be downloaded at http://www.changeover.com/soptemplate.doc.

The example SOP template was developed using Microsoft Word's multilevel list feature. This is a helpful tool for organizing a document. It also allows the checklist corresponding to the SOP to be generated automatically. Subsections and sub-subsections can be added as necessary for optimal organization. This template is organized in 10 sections with Section 10 comprising the checklist. Some companies will wish to add subsections. Others may find some unnecessary and eliminate them. In all cases, once the company has developed a standard template, all sections should always be included. If a section is not required for a certain machine, it should be left in but with a notation of "Not Applicable." This clarifies that they were not omitted in error and maintains a uniform organization across all SOPs for all machines.

STANDARD OPERATING PROCEDURE		
Title:		Page 1 of 1
Written by:	SOP#:	
Approved by:	Rev. #:	
Approval date:		
Effective date:		

1. PURPOSE
2. GENERAL INFORMATION
   2.1   Scope
   2.2   Safety
   2.3   Responsibility
3. MATERIALS
4. TOOLS
5. DEFINITIONS
6. PROCEDURE
   6.1
      6.1.1
         6.1.1.1
7. DOCUMENTATION
8. ATTACHMENTS
   8.1   Machine settings
   8.2   Machine layout drawings
9. REFERENCES
10. CHECKLIST

At the top of each page is a header block. This header block contains certain standard information that will appear on every page of the SOP:

Title: This is the title of the SOP. It should be descriptive but brief; for example, Changeover—Press #1

Page number of total: This provides both the page number being viewed as well as a count of the total pages in the document. Page numbers and page totals should be calculated automatically by the word processing program.

Written by: This is the name of the person who actually typed the SOP. If there are any questions about what was meant, this person can be sought out and questioned.

SOP and revision numbers: Numbering the SOPs can be a great aid in keeping them organized. Various formats can be used, such as breaking the numbers into blocks so that one block can identify the SOP as changeover, cleaning, operating, documentation, or other major functions. In a small plant in a nonregulated industry, this may not be required. In most cases this is a useful feature to help track the SOPs and keep them organized.

Approved by: It is always a good idea to have someone review all SOPs before putting them into use. This approval should be done with an eye toward technical correctness as well as organization and readability. If multiple levels of approval are required, additional blocks can be added to the header template.

Approval and effective dates: These dates tell users when the SOP was put into use. It is strongly recommended that a schedule for periodic review and reapproval be established. This is necessary to assure that what is actually being done on the floor is still what the SOP says is supposed to be done. If desired, a block for a review date can be added to the template.

A "Printed At" date/time stamp can be added. Any time a copy of the SOP is printed, this will print the current date and time. This is particularly useful if, rather than using file copies of the SOP, new copies are printed at time of use. Some companies go even further and offset the "Printed At" by 24 hours, calling it an expiration date. This helps assure that the most current copy is being used.

The sections below the header block describe the various parts of the SOP.

Section 1 Purpose: This section begins the SOP with a statement of purpose. What does this SOP accomplish? In this example it is to provide a changeover procedure for an Acme Labeling Machine. This purpose should be no more than 2–3 sentences long. Shorter is generally better, but take care not to make it any shorter than necessary.

Section 2 General information: includes several items:

Section 2.1 Scope: Scope describes specifically what is covered and under what circumstances. There might be several changeover SOPs for a given machine depending on whether the changeover is container changeover, product changeover, or both. Additionally, there may be several machines in the plant that, though similar, are different enough that each requires its own SOP. Scope will detail which ones are covered. In order to further reduce the possibility of confusion, it is desirable to note when similar machines are not covered by the SOP; for example, "This SOP provides a changeover procedure for the Acme Model A machines located on Lines 1 and 3. It does not apply to the Acme Model A machine located on Line 2."

Section 2.2 Safety: Safety must always be expressly addressed. This may be a general safety requirement such as safety glasses or it might be a more specific reminder to use special protective gear or precautions. This section may also reference safety precautions from other SOPs—for example, "See SOP XXX.XX for detailed lockout/tagout procedures."

Section 2.3 Responsibility: This section spells out who, by job description, is normally responsible for performing the changeover. It may be a mechanic, an electrician, an operator, or some combination of these and other skills.

Section 3 Materials: This section lists all parts and materials necessary for the changeover. This will include cleaning supplies such as sponges or wiping rags, or sample parts or containers required for setup. It will also list all machine parts, that is, changeparts required for the changeover.

Section 4 Tools: This section lists all tools normally required during changeover. This includes wrenches, screwdrivers, and the like, but also includes any specialized tools or gauges that may be required. It must be specific—for example, "½-inch end wrench and ¾-inch box wrench" rather than just "wrenches."

Section 5 Definitions: In some cases, machine nomenclature may not be clear and special definitions may be required. Acronyms and abbreviations should be defined in this section whenever there is any lack of clarity.

Section 6 Procedure: This is the meat of the SOP. This is the detailed description of exactly what is to be done and how. It must be very detailed so that nothing is left to guesswork or judgment. Ideally, it should be detailed enough so that anyone, with no special training, can walk through it step by step and successfully perform the changeover.

In the example, Section 6 is organized in two sublevels. Additional sublevels may be used if desirable. This should be done with caution because too many sublevels may obfuscate rather than clarify. The organization of the sections permits automatic checklist generation.

Section 6 is a first-level item. Section 6.1 is a second-level item and should be a brief description of what is to be done. It should be no longer than 2 sentences and 1 sentence is preferred. A typical Section 6.1 instruction might read "Remove the infeed timing screw." When writing the second-level items, bear in mind that this is exactly what will appear in the checklist.

This simple instruction will tell the experienced person what to do. If the person has been properly trained, this may be enough. On the other hand, this simple instruction may not be adequate for the person who has little experience or who normally does not perform this changeover and who needs additional instructions. If second-level instructions can be thought of as telling the teammate what to do, third-level instructions explain how to do it. The procedure section of the SOP might read as follows:

6.1 Remove the infeed timing screw.

   6.1.1 Manually rotate the machine until the setscrew on the infeed timing screw is on top and accessible.

   6.1.2 Loosen but do not remove the setscrew using a 3/16-inch Allen wrench.

   6.1.3 Slide the timing screw off of its mounting shaft.

   6.1.4 Set the timing screw in its designated position on the parts cart. An annotated picture should be included in this section showing the location of the timing screw, the location of the setscrew, and the direction in which the timing screw is to be slid off of the shaft.

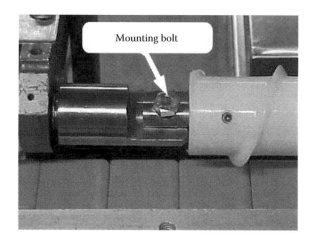

6.2 [Next step…]

The description in Section 6.1 will allow virtually anyone to safely and effectively remove the timing screw. The combined mention of the specific wrench to be used, along with listing that specific wrench in Section 4 under "Tools," helps ensure that the proper tool will be available and used. This avoids the time lost from not having the tool immediately at hand as well as potential damage caused by using the wrong tool.

Section 6 continues with as many subsections and as much verbiage as required to fully describe all of the changeover tasks. It is difficult to make it too detailed. The goal is to put all required knowledge in the document rather than trusting it to be in someone's head. A picture is worth a thousand words and annotated photographs and drawings should be used extensively. It is difficult to overuse them.

Section 7 Documentation: In many cases there will be a log book or other documentation that records each changeover as well as any issues that may have been noticed during the process. If there is not, there should be. This section is used to describe how to complete the logbook and any other documentation.

Cleaning Is Inspecting

Changeover by its nature puts the teammate in intimate contact with the machine. They should be trained to note any discrepancies during the changeover and a means must be provided to report them. An operator, during cleaning, may notice black rubber particles in the area of the drive belts. They may not know where these are from or even whether it is a problem. Nor do they need to know, though it would be helpful if they did. All they need to know is that this is not a normal condition and they must report it as promptly as possible. This needs to be something more than just mentioning it to the supervisor, who probably has many other things on his mind. There needs to be something more tangible, whether it is a formal maintenance work order, an entry in a "Gripe Book," or an informal discrepancy reporting form. Failure to do this early will result in the problem being uncorrected until a minor problem becomes a major failure.

Section 8 Attachments: There often need to be additional documents included with the SOP. One common document is a machine setup sheet. It will likely be cumbersome to put all the individual setpoints into the body of the SOP. The setup sheet will list the height setpoint, temperatures, pressures, timer settings, and more for each product. This may be done with an individual sheet for each product. Alternately, it can be a spreadsheet or table with each product listed across the top, the setpoint name down the left-hand column and the setting for each product in the appropriate column. Where changeparts are used, these can be added to the setup sheet with the part name in the left column and the specific part ID in the appropriate cell under the product name.

The advantage to dedicated sheets is that, assuming the right sheet is used in the first place, there is less chance of using a wrong value based on misreading the matrix. The disadvantage is more pages, but if the original is stored on a computer, this is less of an objection.

Changeover Guideline		
Title: Capper Changeover		Page X of Y
Written by:	Date:	

Printed at XX:XX mmddyyy

CHANGEPART/ADJUSTMENT MATRIX

Bottle Setting	12oz	20oz	36oz	42ml	75ml	100cc	140cc		
Spindle gap	2.5"	2.5"	3.2"	3.2"	3.6"	3.6"	3.6"		
Spindle height	4.0"	4.3"	4.3"	4.5"	4.7"	4.8"	5.1"		
Spindle speed	7	7	7	7	7	7	7		
Side belt speed	6	6	6	6	6	6	6		

Section 9 References: References are additional sources of information that may be useful in understanding and applying the SOP but do not need to be part of it. These can include operating and maintenance manuals, electrical schematics, parts lists, and programming code.

Section 10 Checklist: The SOP for a complex changeover, properly detailed, may be 20–30 pages or even longer. It is not reasonable to expect that an SOP of this length will be used on a daily basis. As they acquire experience, teammates will start working from memory rather than from the SOP. When they start working from memory, they are almost guaranteed to start deviating. They need to have a concise document to assure that they complete all the tasks in the proper order. That is the purpose of the checklist. The checklist tells them what to do.

Microsoft Word has a number of powerful documentation tools. One of these is the Multi-Level List/Outline tool that was discussed above. Another is the Table of Contents function. This function allows for selective listing of sections of the document; in this template, the second level of Section 6. This selective listing, located in Section 10, provides the checklist, for example, "6.1 Remove the timing screw." The table of contents also shows the page number for each entry. This allows the person who is unfamiliar with how to remove the timing screw to turn directly to the page where the detailed instructions (6.1.1, 6.1.2, etc.) are found. Hyperlinks are automatically created from the checklist/table of contents to the page on which the detailed instructions are found. If the document is being viewed electronically, on a tablet or PC, this allows jumping to the instructions in a single click.

In this day of digital cameras and high-quality color printers, there is no excuse not to have SOPs profusely illustrated. It is much better to have too many than too few pictures. Pictures should be annotated, pointing out critical areas, showing directions of movement or rotation, defining part names, and identifying disassembly/assembly sequences as appropriate. Pictures in the SOP should be large enough to be easily legible.

Graphical SOPs

Another approach to checklists is more graphical. A schematic of the machine is drawn showing all adjustment and changepart points. Depending on the machine and changeover process, a well-annotated photo may be used instead. Each point is numbered in the sequence in which it is to be performed. Separate sheets, clearly identified, are developed for each product.

Under the machine schematic, there will be a table showing the location/sequence number from the schematic and the setpoint or the part to be replaced. The teammate, when assigned to perform the changeover, is given the appropriate sheet and works through it, step by step. For critical changeovers, blocks for initials, verification, and signatures can be provided for each step or for the setup as a whole.

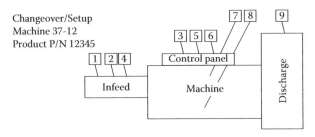

Changeover/Setup
Machine 37-12
Product P/N 12345

Perform all steps in sequence

Setting location/name	Setpoint/Action		Setting location/name	Setpoint/Action
1 Conveyor guide	3703		8 Holding block "B"	Replace with P/N 732B
2 Infeed height	4"		9 Discharge guides	2937
3 Conveyor speed	75FPM			
4 Infeed guide	Replace with P/N 123			
5 Heater Temperature	325 degrees			
6 Machine speed	30 cycles/min			
7 Holding block "A"	Replace with P/N 732A			

On some setups it may not always be possible to make the machine run optimally at the specified setpoint. There may be some variation in product requiring some slight changes to the specified setting. It is important not to write oneself into a corner on an SOP. Setpoints generally should be described as targets and some minor variation allowed at the team member's discretion. If parts come in with a bit more flash than normal, it might be necessary to open a guide slightly to prevent excessive jamming. If the SOP does not make allowance for this, the team member may comply with the setpoint and risk subpar performance. Alternately, they may open the guide slightly. In a regulated industry, such as pharmaceutical, a government inspector may note that the guide is not at its stated setpoint and make a "not following own SOPs" observation.

Variations from the target setpoint should be recorded. If it is a onetime occurrence, it is not necessary to take any action. If it is found that the deviation is routine, the target setpoint should be reviewed and updated if necessary.

Writing the SOP

After developing a standard format for SOPs, it is time to begin writing them. If they do not currently exist and are to be written from scratch, this may seem like a daunting task. There are a couple of ways to ease into this.

Writing SOPs takes some practice and experience. It is best to start with simpler changeovers and work up to more complex changeovers. This will make the inevitable mistakes more obvious and thus more likely to be corrected. It is also likely to make the mistakes less serious.

SOPs should be written in the early stages of the changeover improvement program. One issue with improving changeover can be that nobody really understands what the current practice should be or even is. Writing SOPs at an early stage will document current practices. Once documented, these can serve as a starting point for improvement. It may be desirable at this stage to bypass all the formal approvals and recognize these as working documents on the way to formal SOPs. Rather than calling them standard operating procedures/SOPs, which can have legal and regulatory implications, they can be called *changeover guidelines*.

A good starting point for developing a changeover SOP is the equipment manufacturer's manual. Many of these will include instructions on how to perform an SOP as well as setpoints developed during equipment testing. When purchasing new equipment, this should always be insisted upon. These are a good starting point, but are often written somewhat generically. Setpoints developed during testing may not always reflect actual conditions in the field. When machines have been in use for a while, they may have been modified in the plant. These modifications should have been noted in the machine manuals as they are made, but they may not have been.

There is an easy way to test this. Give a team member the manual and ask this person to perform a changeover from it. Be sure to record all comments, uncertainties, and deviations from setpoint required to complete the changeover.

Another way to develop an SOP is to appoint a writing team. This team will consist of a person or persons who carry out the normal changeover process and a second person who will record what they are doing. Both members should be carefully chosen. The person doing the changeover must be someone who is experienced and recognized as doing the work correctly. The recorder (writer) should be someone with writing skills who can clearly describe what is being done. Some knowledge of the changeover process may be helpful, but is not strictly necessary. However, they must recognize what they do not know and not be shy about asking questions. Correct nomenclature at this stage is very important. It may be found that different people performing the changeover use different names for some of the components or adjustments. The recorder must determine which is the most correct, perhaps with the assistance of machinery manuals, and incorporate that into the SOP.

On larger machines with more than one person working on the changeover, it may be hard for a single recorder to capture everything that is being done. Even when a single person is performing the changeover, that person may still work faster than the tasks can be written down. A video camera can be very helpful in documenting the changeover. This does not need to be an artistic masterpiece, but it does need to show at least an overview of what is being done. It is generally sufficient to mount the camera on a tripod with a good general view of the changeover, start recording, and let it go. The video camera does not relieve the recorder of the need to take copious notes. Its purpose is to allow the person to go back over the changeover to assure that nothing was missed. If necessary, when reviewing the video, the person who was doing the changeover can be consulted for clarification and more detail.

After recording the current changeover practices, it is a good idea to have them reviewed for accuracy by the teammates currently doing changeovers.

This will reveal discrepancies about how tasks are to be performed as well as discrepancies about various component names. One useful technique is to get everyone in a conference room and project the SOP on a screen. The group can then review, line by line, until an agreed upon description of the current practice is established. This group approach takes advantage of the group's collective knowledge and allows differences of nomenclature and opinion to be resolved. More importantly, it gives the teammates direct input into the document as well as experience working as a team. Both enhance a sense of ownership.

Once existing practices are documented, they need to be verified. This is done by providing a copy of the SOP to a person normally performing the changeover. This person should then perform the changeover exactly as documented. Where the SOP is incorrect, this must be noted and revised.

After verification, the SOP should be put in use. This means that everyone should be trained on it as the only way that changeover is to be performed. Supervisors and managers must do their jobs in ensuring that the documented practices are followed to the letter. Standardization is an important first step in improving changeover. Standardization is likely to reduce changeover time. It is especially likely to reduce startup time. Even where it reduces neither, it will make both less variable, which will aid in planning and scheduling machine capacity.

The is only the beginning, not the end of the process. As the existing practices are documented, it is likely that opportunities for improvement will become apparent. As they are noted, they should be noted. Some may be simple and can be addressed immediately with the improvement incorporated. Others may be more difficult but should be noted so that they can be addressed at a later date. There must be a process in place by which SOPs can be modified as errors are discovered, conditions change, and improvements are implemented. This process must be more formal than simply changing the SOP willy-nilly. On the other hand, it must be flexible enough to allow changes to be made relatively easily and promptly. Failure to provide a good revision mechanism can lead to situations where the SOP does not match the methods used. Particularly in a regulated industry such as food or pharmaceutical, this can have serious consequences. The single largest cause of observations arising out of Food and Drug Administration (FDA) inspections is failure to follow one's own procedures. Not having an SOP may be viewed as a less serious offense than having one and not following it.

Electronic SOPs

An important factor to be considered is the conflict between control and accessibility of the SOPs. On the one hand, it is important that the team-mates who are expected to follow the SOPs have access to them. No matter how good their training, unless they have the actual SOP available when performing the changeover, they will deviate from it. This can mean having a number of copies; at one extreme, one for each teammate. When it comes time to revise the SOP, recovering all of the previous copies and replacing them with the new can be a cumbersome task.

The other extreme is to have a limited number of copies. This makes them much easier to control and to assure that all are always current. At the extreme, this may mean only a few copies kept in the offices of supervisors and managers.

Neither extreme is desirable; a balance must be struck between access and control. When SOPs are not readily accessible, teammates will find workarounds that permit them to do their jobs. One workaround will be to make bootleg photocopies for individual use. This must not be allowed. As hard as it is to collect authorized copies for replacement, it is several orders of magnitude more difficult to collect the bootleg copies. Strict enforcement against extra copies is required. One tool that may be helpful is to print the authorized copies in a different color ink. Another is to stamp the copies with a red or other distinctive color stamp. Any copy found in black or with a black stamp will be unauthorized and must be confiscated. The teammate must also be trained why they cannot have it.

Electronic systems can give the best of both worlds. One early implemen-tation in the 1990s was at a company with a half-dozen plants around the United States. SOPs were developed and maintained by the individual plants, but the originals resided on a server at the corporate offices. They could be read on the screen or a copy could be printed. If a copy was printed, a time and date stamp was automatically included at the top of each page. This indicated that the copy expired 24 hours after printing.

Anyone with appropriate network access could access the SOP. Local termi-nals and printers on the plant floor assured that each teammate had easy access.

In recent years, as enterprise systems such as SAP have been integrated into manufacturing, this type of system has become more common, though perhaps not as common as it should be.

Electronic SOPs can bring their own problems, of course. It is critical that appropriate security precautions be taken to prevent unauthorized changes. Tracking systems must keep track of who accesses the system, what they did, and when they did it to provide an audit trail in the event of problems.

A new area that seems to open endless possibilities is the use of tablet computers as well as other similar devices on the plant floor. The Apple iPad may be the most obvious example, but there are other options as well. Smaller tablets, including Amazon's Kindle Fire, Android and iPhones, MP3 players including the iPod, netbooks, and even personal digital assistants (PDAs) offer some intriguing possibilities.

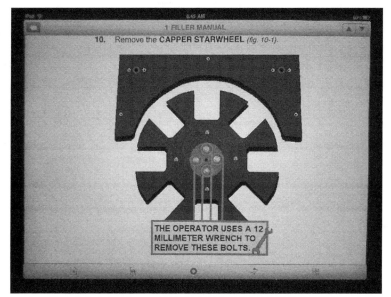

Photo courtesy of Nail Creek Services Inc.

All of the above are Wi-Fi capable, so there is little issue with memory capacity on the device itself. The SOP can reside on a central server, which resolves the control issue. The device can allow printing of auto-expiring copies for those who prefer to have a paper copy in their hand. In that sense they are similar to the system described above.

There are a number of additional benefits that can't be obtained from a paper copy. First is hyperlinking. When the checklist refers back to the SOP section for elaboration in detail, the teammate has only to touch the screen to be hyperlinked immediately to the appropriate section. This can also be linked to other SOPs. If one task is to lock/tag out the equipment, the SOP might say "lock/tag out the equipment in accordance with SOP XXX." Making this a hyperlink, the teammate can be jumped to the specific uniform lock-/tag-out SOP.

Another benefit is electronic checklists. This allows the teammate to touch a checkbox beside each task indicating completion. This provides a real-time record of what they have done and have left to do. It also provides a permanent record showing that all steps were performed and at what time. If it is desirable to prevent one step from being displayed until the previous one is ticked off, this too is possible.

These devices put an almost infinite virtual library in the teammate's pocket. Machine manuals, drawings, logic diagrams, tool-room parts availability, and other information, once in electronic form, can be accessible as needed. This prevents lost time looking for information as well as misplaced or damaged manuals.

Depending on device and machine capabilities, it may be possible to program the machine to automatically send text messages when abnormal conditions such as overheating occur or when maintenance is due. This can aid in assuring more timely intervention.

Finally, TV. The audio and video capabilities of these devices open up a world of possibilities. This can include fairly standard-type training videos, but accessible wherever and whenever needed. These can be taken even further and made interactive using either video, animation, or a combination. In one demonstration the author attended, a mechanic performing a setup wore an iPod strapped to his upper arm. Bluetooth earphones gave him the audio. The iPod walked him step by step through the machine setup, pausing frequently. As the mechanic grasped each step, he could move on to the next. If uncertain, he could replay any prior step. The arm-mounted iPod was out of his way, allowing free movement to perform his work, while almost never leaving his peripheral vision.

Several equipment manufacturers have begun mounting webcams connected to the user's network at strategic points in their equipment. Some are stationary, focused on key components. Others can be panned and zoomed remotely. These cameras allow a service technician in New York to view and diagnose problems with a machine in California without needing to go personally. This same concept can be adapted in plant to allow technicians to remotely view machines and take necessary action without needing to go to the machine first.

Conclusion

Some people may feel that SOPs are difficult to write, hard to control and maintain, and take away from a technician's professional expertise. They are right to some degree about all three. Consider the alternative, though. It is impossible to perform consistent changeovers without written instructions. If changeovers are inconsistent, equipment operation will be inconsistent as well, especially at the start of the production run. Inconsistent production will result in inconsistent products at the end of the process. It may be that most or even all products are within specification. It may be that products that are not in specification can be caught and rejected. The end result is that the customer is getting variation, and the more variation they get in the products they buy, the less they will like it.

Good, detailed SOPs, properly written and rigidly followed, are the key step in driving out this variation.

Chapter 4

Eliminate

The ESEE approach to lean changeover (LCO) has four components: Eliminate, Simplify, Externalize, Execute. The most important of these, by far, and the one that must be done as the first priority, is elimination. It is a double waste of time to improve something that should not be done in the first place. It wastes the time spent on improvement and it wastes the time spent every day carrying out the, albeit improved, task.

A manufacturing process may be likened to a family Christmas tree. In the beginning, there are not a lot of ornaments to hang on the tree. Over the years, more and more ornaments are acquired, and because they are available, they are hung on the tree. At some point the tree becomes over-burdened with decorations and a weeding out is necessary.

Manufacturing processes tend to start out fairly lean with only what is truly necessary. Over time they accumulate many "ornaments," which may have served a purpose at one time but are no longer adding value. The first step in any LCO program must be to identify and eliminate these. Elimination opportunities can occur in the following areas.

Product Design

Product and component design can provide significant opportunities for improvement. One medical device company manufactured products in 5- and 10-milliliter glass vials. Visually, their diameters appeared similar. In reality, there was a 0.063-inch difference. This does not seem like a lot, but it required a 3-dimensional changeover to each machine in the line. This took 90 minutes.

Had the vials been the same diameter, there would have been a height adjustment required to the filler, stopper applicator, and the overcapper, taking about 5 minutes. The wasted 85 minutes per daily changeover represented about 20% of total plant capacity.

10 ml

5 ml

Most suppliers design families of vials to avoid these issues. It would have been a simple matter initially to specify two vials of the same diameter. Due to regulatory requirements, changing it later was impossible as a practical matter.

A consumer goods manufacturer produced over 800 stock-keeping units (SKUs) on cardboard hang cards and faced two product design issues. Each caused significant excess changeover time. At some point someone changed the size of the card. All were the same width, but about half were 1/8 inch longer than the others. They changed products as often as 4–5 times per shift and frequently had to adjust the assembly machine to the different card sizes. This was not a major task and took about 5 minutes. While this may not seem like much, it added up quickly. Five minutes repeated 4 times per shift represents 45 shifts of lost production over the course of a year.

5 minutes/changeover × 4 changeover/shift × 3 shifts ×
360 days = 21,600 minutes/year = 45 shifts/year

A second issue with this product was that each card could carry 4, 6, or 8 individual products. There was little problem changing between 4 and 8 products. It was a matter of blocking or opening 4 of the 8 lanes that normally fed the product. Changing to and from the 6 pack was more difficult and required reconfiguration of the lanes.

The author worked with plant management to document the lost time caused by these two design issues. The case was presented to marketing, which was responsible for package design. They realized that the two card sizes were creating problems, but were unaware of the magnitude. Once they understood, they immediately began a program to standardize all cards to one size.

The issue of the blister was more difficult. They felt that some customers would feel that 4 products were not enough but that 8 were too many and that they needed to continue offering a 6 pack.

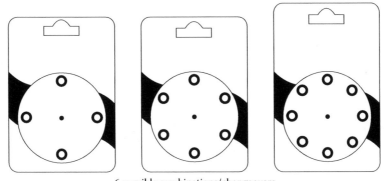

6 possible combinations/changeovers

This is an important example of the old expression "The customer is always right." It is the plant's job to manufacture what the customer wants, no matter how difficult. It is not the customer's job to change their preferences to make the plant's job easier. Changes to product design to facilitate changeover must always be made with the customer's wants foremost. In the two previous examples, neither change impacted the customer's experience with the product.

Many companies run a multiplicity of product sizes. If not required for marketing or competitive reasons, it may be possible to reduce the number of sizes. One company produced pharmaceutical tablets in 11 different sizes of bottles. They were able to combine some similar size bottles. Rather than have both a 30- and a 50-cc bottle, they eliminated the 30-cc version. The 50-cc bottle was a bit oversized and wasteful for the product that had run in 30 cc. However, eliminating the need for changeover and eliminating the smaller bottle as an inventory item more than compensated for this.

Another possibility is to permit a wide range of bottle dimensions and shapes, but to standardize "touch points" and overall height. *Touch points* are those points on a bottle where the machinery contacts it. On a 20-ounce bottle for carbonated beverages, this can be just below the neck and about

1 inch above the base. At any other point on the bottle, diameter and even shape can vary as needed. As long as the two touch points and height are held to a common dimension, little or no changeover between bottles will be required. One trend is to handle certain types of beverage bottles solely by their necks. As the necks are identical for all bottle sizes, this allows any size to be accommodated from the infeed through the rinser, filler, and capper to the conveyor or infeed with no changeover adjustment.

Documentation

Record keeping and documentation can offer opportunities for elimination. Henry Ford addressed this issue 90 years ago in his autobiography *My Life and Work,* saying:

> We abolished every order blank and every form of statistics that did not directly aid in the production of a car. We had been collecting tons of statistics because they were interesting. But statistics will not construct automobiles—so out they went.

Some documentation will be required for legal or regulatory reasons and cannot be eliminated. Other documentation may have been necessary in years past but no longer is. If not, and if it continues to be collected mainly out of inertia, it must be discontinued. Some plants duplicate documentation. There may be a warehouse order, manufacturing order, and quality record. Much of the information—sometimes the majority of the information—on the various forms may be duplicated. It may be possible, especially if using computerized enterprise requirements planning (ERP) systems, to develop a single document with information from the various departments linked to eliminate duplication of effort in filling and maintain the forms.

The entire documentation cycle needs to be examined from a zero base. Every form, every data entry, must be justified. Unnecessary collection of data is not only wasted time, it can potentially have other harmful impacts down the road. Collect what is truly required and no more.

Quality Clearance

Some production lines or processes must be inspected or "cleared" by quality at some point. This may be at the beginning, at some milestone, or at the end of changeover. This clearance itself can take some time. In addition to the time taken for the clearance, a great deal of time may be wasted in some cases waiting for the quality inspector to arrive. Better coordination between departments can help this, but even better is eliminating the operation. Some plants use the same operators who clean and set up the line to perform the clearance. There are reasons for and against this, apart from reducing changeover time, and they all need to be examined carefully to assure that product integrity is never jeopardized.

One converter of paper and plastic films required testing of the static elimination devices before production runs. This was a safety requirement as static buildup created a fire hazard. It could not be eliminated but did cause two avoidable production losses.

First, the test had to be performed by maintenance. Once the operators had finished the changeover, they would call maintenance. After some waiting, the technician would arrive to perform the test. There is a huge disparity in cost between the technician and machine downtime. Waiting should always be minimized, of course, but if there is to be any waiting, it must be the technician waiting on the machine, never the machine waiting on the technician. This was solved by calling maintenance at the beginning of changeover rather than on completion.

The second issue was that the coating could not be mixed until the static testing was completed. The coating had a short shelf life after mixing and the worry was that if the static test failed, the delay in startup would cause loss of the coating if mixed before. If costs are known, it is easy to determine the potential cost and benefit of mixing ahead of time. It is the cost of the coating, adjusted by the (low) probability of static test failure compared to the cost of 30–40 minutes of additional downtime every time the product is run.

Schedule Production Weak to Strong

There are many different techniques for determining production schedules. Some take changeover issues into account. All should. There may be opportunities to eliminate or simplify changeovers through improved scheduling practices. One such practice is sequencing. A paint plant will normally produce a variety of colors. If production can be scheduled light color to dark color, cleaning between colors can be minimized or possibly eliminated. If changing from red to white paint, a thorough cleaning is required. Even small traces of red paint remaining in the system will give the white paint an off color. If changing from white to red, some pigging or flushing may be all that is required. Trace amounts of white paint remaining in the system will not be sufficient to discolor the red paint.

Similar applications include pharmaceutical packaging, going from weaker to stronger dosage of the same product with minor cleaning between them. A food plant may be able to shift from nut-free products to products with nuts with little or no cleaning during the week. At the end of the week a major cleaning is performed to allow nut-free products to be run again. Juice and water bottling plants can run the water first because small amounts of water will not harm the juice but even trace amounts of juice will give the water an off flavor.

Breweries often produce various different beers and rather than cleaning the pipelines, they may flush to drain as they change from one beer to another. This saves time by eliminating cleaning but does so at the cost of wasted product. The decision on which path to choose can be made by comparing costs. The cost of flushing 100 gallons of beer is perhaps $200, including both product and disposal costs. Assume that the cost of downtime is $12,000/hour and cleaning will cause 20 minutes of downtime. The cost justification is simple, $200 for flushing versus $4,000 for cleaning. This is fairly easy to see with beer because of the relatively low cost of the product and the high speed of the bottling line. The same evaluation technique will work with a line producing spaghetti sauce but the answer may be less obvious.

Building Design

Building or facility can impact changeover positively or negatively in a number of ways, starting with size. On the one hand, production areas should be as large as necessary, perhaps with some extra space for future expansion or improvements. On the other hand, the larger the area, the more time will be spent in cleaning. Excess floor space can also lead to a tendency to accumulate more things in the room than are absolutely required, including work in process (WIP). High ceilings may be nice but also mean more wall surface requiring cleaning. Higher walls will not only take more time, the extra height will make it more difficult.

Materials of construction for floors, walls, and ceilings must be chosen for ease of cleaning. Piping, conduits, and other utilities should be run inside the walls or ceiling so that they do not become dirt catchers in the production room. Where this is not feasible, chases should be provided to cover them.

Separate Clean and Dirty

Some processes are dusty or dirty (dirty in this case only meaning that they require more cleaning). Packing a powdered food product in bottles is naturally dusty. Rather than put the entire line in a single room, the filling and capping operations should be enclosed in a small room with the rest of the line in another area. This isolates the dusty operation and the dust in a more easily cleaned area and eliminates much of the cleaning requirement for the balance of the line.

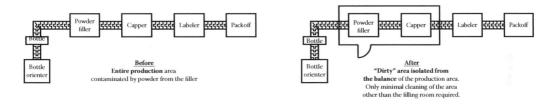

Don't Clean Unused Equipment

Some lines may have equipment mounted that is only used occasionally. When not in use, the machine is left inline and product routed through or around it. If it is left in the line, especially for food or pharmaceutical production, it will usually be necessary to clean it between production runs, even though it is not used. This cleaning can be eliminated by making the machine portable. Mount it on castors and design a docking station. The docking station consists of some mating guides on the movable machine and on the fixed line. When the machine is needed, it is rolled into the area, mated to the line, and latched in place. Plug connections for power, controls, and pneumatics are all provided at the docking station. When not in use, it is disconnected and removed from the area.

Batch Versus Continuous Processing

In some cases it may be possible to eliminate changeover by redesigning the production process or using new technologies. One manufacturer produced a cosmetic product in a wide variety of color shades. Previously they would produce a batch of a particular shade in a tank. The tank would be wheeled to the line and the product filled into jars. The line would then be cleaned for the next shade of product. They were able to convert from batch processing to an inline blending system. This allows them to have a universal base component that is the same neutral color for all product shades. Just before the base is filled into the jar, colorant is blended in to make the final desired shade. This allows them to change shades on the fly and eliminates the entire batch mixing process.

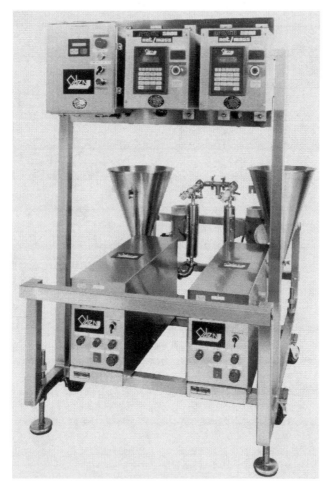

Photo courtesy of Oden Corporation

Label Position

Even something as simple as the position of the label on a shipping case can have significant impact.

One company shipped their products in several sizes of corrugated shipping cases. These cases had labels applied to them with product information and barcodes. These labels were centered on the side of the case, which required resetting the labeler every time the case size changed. This setting was not difficult but did require about 5 minutes (5 minutes/day = 20 hours of lost production per year) to reposition the photoeye and raise or lower the labeler head.

The label was repositioned from the case center to the leading lower corner, approximately 1 inch from the bottom and 1 inch from the edge. As this position was the same for all case sizes, the need for labeler adjustment was eliminated.

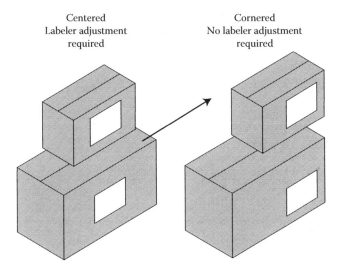

Another opportunity can arise in positioning labels on bottles. If the labels are centered vertically on the bottle, this will require resetting the labeler head for each bottle change. If the labels can be repositioned so that the bottom edge is always the same distance from the bottom of the bottle, the height of the head never has to change.

In another instance, due to changes in the width of the label between bottle sizes, the position of the lot and date codes, printed during label dispensing, varied between products. This variation required repositioning of the imprinter with each label size change. This was not a major task, requiring only a few minutes, but when the code position changed, the camera that inspected for code presence and correctness had to be repositioned, which was a big issue. This was a delicate and difficult adjustment and required a technician anywhere from 15–30 minutes to perform. Standardizing the code position on the label relative to the bottom of the bottle would have eliminated the need to reposition the coder and camera.

In this particular case, it was agreed that it was a good idea, but because it was a pharmaceutical product, the regulatory, approval, art changes, and other issues made it impractical. This is another example of a change that would have been simple to carry out in the design stage of the labels but became essentially impossible to do retroactively.

Cam

There will be opportunities to redesign machines to eliminate changeover tasks. One food plant ran several lengths of pouch. Length was determined by the position of a cutoff shear activated by a cam on the machine main shaft. When changing pouch lengths, a mechanic would move the shear to the proper position, then open the lower machine cabinet and adjust the position of the cam on the main shaft. This was time consuming, and because it required tools, had to be performed by a mechanic rather than an operator. One of the mechanics decided that this was unnecessary. He had a machine shop fabricate a cam with an identical profile but about 3 inches wide versus the original ½-inch width. This width was sufficient to drive the shear arm for any pouch size from shortest to longest with no repositioning of the cam. As part of this project, the bolt fixing the shear arm in position was replaced with a hand lever, and three positioning stops, one for each pouch size, were added. This completely eliminated the need for a mechanic to position the shear. The operator now loosens the hand lever, moves the arm to the appropriate stop, and retightens the hand lever.

Ladders

For safety reasons, every effort must always be made to avoid the use of ladders. Ideally, all changeover tasks should be done from floor level. This is not always possible, and when required, proper ladders must be available and stored as close to the point of use as possible. Mechanics and operators may feel that the effort required to fetch a ladder is more trouble than it is worth. Or, they may feel constrained by time pressure. Whatever the reason, it is not uncommon to see teammates climbing on pipes, control cabinets, or other parts of the machine to gain access. This is unsafe and will lead to falls as well as damage to the machine. They may even use a chair, forklift, or something else instead of a proper ladder.

In some cases it may be impossible to completely eliminate the need to climb onto a machine to reach adjustments. One alternative is to mount commercially available bolt-on steps to the machine. These fold up and out of the way when not in use.

In other cases, these may not be sufficient. One option is to weld cross-members to the machine frame, making a fixed ladder. These should have nonskid tape on them to indicate that they may be used as a step as well as to reduce the risk of slips and falls.

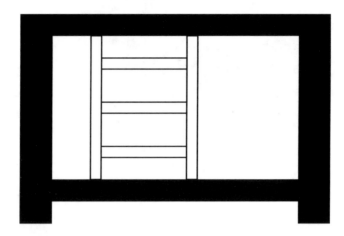

In some cases, it may be useful to add a small platform to stand on when performing changeover. This platform may be fixed or may be on spring-loaded castors so that it sits directly on the floor when weight is applied. It can then be moved out of the way when not in use. Another alternative is a platform that can be folded up or down to get it out of the way.

Quick Connectors

It is often necessary to break and make hose connections during change-over. These may be small tubing for actuation of pneumatic actuators or may be larger hoses for cooling water to molds or other components. Lightweight large-diameter hoses are commonly used for dust collection. Whatever the hose, they should always have quick connect, tool-less, fittings. In many cases the sequence of the hose connection is critical. Pneumatic cylinders will have a connection at each end. Dies will have cooling water in at one end and out at the other. When hoses are connected backwards, severe damage to the machine may result. One solution is to use multiple hose connectors. These combine several tubes or hoses into a single quick-connect fitting. This combination eliminates the need for multiple connections as well as the possibility of improper connection. This mistake-proofing may eliminate the need for a mechanic, allowing operators to make the connection while the mechanic is freed to perform more valuable tasks.

Multiple connections for larger hoses, such as cooling water or air or hydraulics are also feasible.

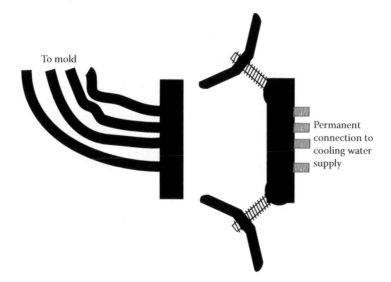

When hoses must be fitted over hose barbs or where a large-diameter dust collection hose must be fitted over ducting, worm-style hose clamps are commonly used. These should be replaced with toggle or other quick-release style clamps to eliminate the need for tools.

Pucks

It may be possible in some circumstances to completely eliminate the bulk of changeover. Pucks and pallets can be used on packaging, assembly, and other lines to eliminate the need to change the product handling or positioning components. Pucks and pallets standardize the handling or outside dimensions across a range of products. They also standardize the part or product position, including height and centerline between parts. Pucks and pallets also provide a means of handling parts that would otherwise be difficult, such as a round-bottom bottle or an irregularly shaped molding or stamping.

Pucks are plastic cups that are used to carry individual bottles or other packages through the packaging process. For a given line, they should all have identical outside size and shape. The inside dimensions are varied to fit the bottle. The shape of the puck does not need to be the same shape as the bottle. This allows a square or oval bottle to run in a round puck for ease of handling on conveyors and other machinery. Multiple shapes can also be used. A puck can have a round bottom section for ease of handling. A rectangular upper section allows it to be axially oriented for other operations.

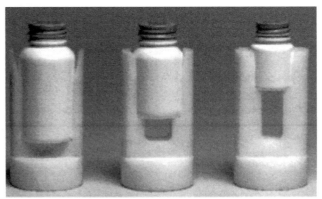

Photo courtesy of TAP Biosystems

Pallets are often aluminum plates to which a part-specific fixture is mounted. All pallets for whatever parts are the same outer dimension, eliminating the need for any adjustment to the handling system. The fixture is designed so that all parts, when mounted on the pallet, have common heights, centerlines, and spacing. Pallets can be rectangular and may have rounded corners or even radiused ends to facilitate their movement through conveyor curves.

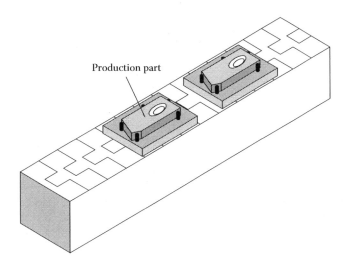

Production part

Pucks and pallets continuously circulate through the system. At the infeed of the line they are loaded, passed through the processing stages, and the finished part removed. The empty puck/pallet is then transported back to the infeed via racetrack or overhead conveyor where another part is loaded for processing. Parts may be placed and removed manually or via automated machinery.

In one extreme example, a company uses pucks to allow 3 bottle sizes, 2 neck sizes, and up to 6 different tablets and/or capsules to be packaged simultaneously at speeds of 60 bottles per minute.

It starts with 3 bottle-orienter or puck-placing machines, 1 for each bottle size. Each puck has a radio frequency identification (RFID) chip embedded with a unique identifying number. As a production order comes to the head of the sequencing queue, a signal is sent releasing 1 bottle/puck onto the line. As this enters the line, its RFID code is read and tied to the production order. The bottle/puck is conveyed to the filling section and the RFID code read. This directs it to one of the 6 filling machines, depending on the requirements of the order, where it is filled with the required number of tablets. If more than 1 type of tablet is to be filled into the same bottle, the sequence is repeated at the appropriate filler. Once filled, the bottle is moved to capping. The RFID code is used to direct the bottle to the right capper. After capping and induction sealing, the bottles are labeled. The labeler is loaded with generic labels that are all the same regardless of product or bottle. As the bottle/puck enters the labeler, the RFID code is read and matched to the production order. This causes a printer integrated into the labeler to print the appropriate product, count, dosage instructions, and other information on the label as it is applied. The bottle/puck joins others going to the same destination. When the order is complete, they are released for hand packing. The pucks are conveyed back to the line infeed where they repeat the process.

A distant cousin to the puck concept has become more common in recent years in the beverage industry. Products packaged in polyethylene terephthalate (PET) or other plastic bottles, such as soft drinks, can have a common neck geometry regardless of overall bottle shape and size. The bottling line is designed to handle bottles by their neck rather than by their base or body. This allows bottles as diverse as 6 ounces and 2 liters to be run with no changeover from the blow molding machine, through conveying and accumulation, rinsing, filling, and capping up to the infeed of the labeling machine. This eliminates any need for changeover of the machines between container sizes.

Servo Motors

In the not too distant past, many production machines had a single relatively large motor. Combinations of belts, chains, gears, cams, and shafts were used to distribute power from the motor to the various machine functions. When it was difficult to transmit the power mechanically, pneumatic or hydraulic cylinders and motors would. These mechanically complex machines could be difficult to keep synchronized. When changing from one size to another, they could be even more difficult to resynchronize. This was partly due to complexity, but also partly due to the nature of mechanical systems. No matter how well designed and built, all mechanical systems will have some play. As they age, this play only gets worse, even with good maintenance.

Another drawback to mechanical systems occurs during changeover when they are adjusted from one product to the next. This is time consuming and imprecise, often requiring a skilled technician to do it properly. Even a skilled technician may not be able to get it exactly right the first time. Chapter 7 of this book will discuss how to improve this process. Servo motors can provide a means to eliminate it.

In recent years as servo motors have become cheaper and more versatile, there has been a trend away from one big motor to multiple smaller motors. Instead of using shafts and gears to drive a flap tucker on a cartoner, the tucker has its own small servo motor.

The use of servo motors allows for considerable mechanical simplification of the machine. Instead of needing to reposition the flat tucker by trial and error, or even by a gauge or scale, it can be repositioned electronically. For each package, all the setup positions are programmed in the main control panel. The operator merely selects the appropriate product from a menu on a display screen, confirms that this is what is desired, and the controller does the rest, repositioning all servo motors to the new requirements. Much of the changeover process, particularly much of the imprecision, is eliminated. Also eliminated is the need for skilled technicians as less-skilled operators are now able to perform changeovers with a button push.

One potential drawback to this type of system is that it requires a higher skill level to program, maintain, and repair than might be required with a comparable mechanical system. If these systems are introduced, it may be necessary to provide training to existing technicians or even hire new ones with the appropriate skill sets. In some locations around the world, this may be difficult and it may be advisable to stick with mechanically based machines that are within the capabilities of available technicians.

Machine Covers

Machine covers must often be removed for changeover. Many machine covers have excess fasteners holding them in place. There may be safety reasons for this in some cases, and a safety specialist must always be consulted before making any changes. In some cases the number and type of fasteners can be eliminated or reduced without compromising safety. Covers are sometimes duplicative and may not even be necessary. In one instance a dust collector had an expanded steel mesh grill protecting the filter. On top of the grill, it also had an aluminum louver that required 6 screws to remove. The louver was necessary mainly for appearance.

Some fasteners may also be eliminated by providing a track or slot for 1 or 2 sides of the cover. The cover is held in place by fasteners on the remaining sides.

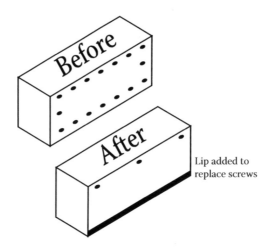

Lip added to replace screws

Eliminate Conveyor Rail Adjustment

Conveyor rails often need to be adjusted to accommodate different product sizes. This is often not critical except at the point that the product enters a machine where it must be precisely positioned. Depending on the product, it may be possible to eliminate much rail adjustment. If the products remain separated on the conveyor with no accumulation, rails serve little or no real purpose other than to keep the occasional bottle from falling off. Where bottles do back up, the required fit of the rails may be a function of the bottle shape. Round products can usually run multiple products across the conveyor without jamming. At a machine entrance, the guide rails can easily return them to single file.

If the products do need to be closely guided by the rails, consider whether both rails need to be adjusted. There is no rule that says products must always be centered on the conveyor. In some instances it may be possible to lock one rail in a fixed position and make all adjustments to the other side. This eliminates half the time for conveyor adjustment. If taking this approach, hand levers should be provided on the adjustable side to eliminate the need for tools. The other side should be fixed in place with setscrews to avoid inadvertent adjustment.

Another elimination tactic is to identify how closely the support brackets need to be spaced. If there are more brackets than required, the excess should be eliminated. One plastic blow molding plant had over 500 linear feet of pneumatic conveyor carrying empty PET bottles between the molding and palletizing area. Guide rails were supported by brackets every 2–3 feet. The bottles were lightweight, carried by their necks, and the main purpose of the rails was to prevent excessive lateral swaying of the bottle. It was determined that half the brackets could be eliminated on the straight sections of the conveyors. In this case, rather than removing them, they were unfastened from the rails and left in place during the next changeover. This eliminated the need to reset over 100 brackets. Since each took about 30 seconds, this represented a savings of 50 minutes of changeover time.

Multiple Photoeyes

Many machines use photoeyes or other sensors to detect product or product features. These sensors frequently require adjustment of position, sensitivity, or both during changeover. This adjustment can sometimes be touchy and time consuming. Some photoeyes are easier to adjust than others, but adjustment, however easy, still takes time and still leaves the possibility of error. These are generally easy adjustments to eliminate. Photoeyes and other sensors are small enough and cheap enough that individual sensors for each product can be permanently mounted. They are connected to the control system via a selector switch. In more sophisticated systems they can be automatically switched by the programmable logic controller (PLC) as part of the setup recipe. This not only eliminates the time and issues caused by adjustment of a single sensor. It also allows the change to be made by an operator rather than taking a mechanic away from more valuable work.

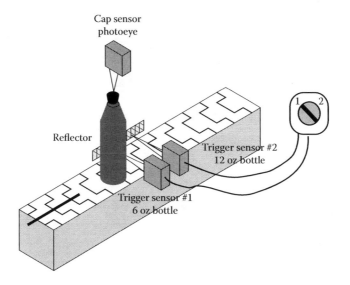

Kit Manufacturing

One company made a product consisting of 4 small bottles, each with a different color. They ran this as many companies do. They would make red on Monday and send it to the warehouse. They would change over and run blue on Tuesday, send it to the warehouse, and change over again. Yellow and green were more of the same. On Friday all 4 colors would be brought out and combined into packages.

They figured that a better way would be to make all 4 colors simultaneously and keep them collated in the 4 color kits. A machine builder modified one of their standard machines to fill and cap 4 colors at a time.

Bottles feed into a starwheel where they index, 4 at a time, into a filling station. There they are stopped and 4 filling nozzles fill a different color into a bottle. When filled, the bottles are indexed to a capping station. There are 4 capping stations, each applying a different color cap. (The cap color plus a generic label is the primary identification.) Another index and the bottles exit the starwheel collated into sets of 4 for packing into cartons.

The value-adding work remained. They still needed to get the product into the bottles and the bottles capped. What was eliminated was the non-value-adding work caused by needing to run, essentially, 5 different products (one each per color plus packing into kits).

Under the old system, the company spent 5 full days a week on production. The kit filling system eliminated enough wasted material handling, cleaning, paperwork, and other busywork that they were able to run their normal production in 2½ days.

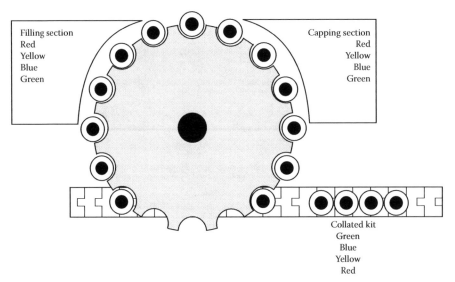

Filling section
Red
Yellow
Blue
Green

Capping section
Red
Yellow
Blue
Green

Collated kit
Green
Blue
Yellow
Red

Preset Timing Adjustments

Many machines used in packaging will use timing screws and starwheels to time the product through the process. In order for the machines to work at all, they need to be synchronized. In order for the machines to work well, they need to be synchronized perfectly. Some machines may have a train of up to 5 or more screws and starwheels that need to be synchronized with the machine itself. This can be difficult to do and is always time consuming. In the best case, it will require a mechanic. In some cases it may be beyond the skills of some of the mechanics, requiring that someone with more experience be diverted from another task.

A common approach is to build timing adjustments into the machine. The parts are mounted and then a mechanic adjusts the timing until it is correct. Often there is pressure on them to get it done and they wind up not taking the care needed. Settings wind up being "good enough" to get the line restarted. Then, with production running, the settings are tinkered with until they are better.

Timing of these parts should only be done once, the first time the parts are run. The drive shafts and hubs should have pins so that when the timing screw or starwheel is mounted it is automatically in the proper position. Any adjustment should be in the part rather than the machine.

The following figure shows a timing screw shaft with drive pin. The timing screw is slid onto the shaft and the pins engage a slot in the screw collar. The first time the screw is run, it will need to be timed. This is done by mounting it and loosening the setscrews holding the collar. The screw is then rotated relative to the collar until synchronized and the setscrews retightened. Once set, this adjustment should never need to be done again. Note the asymmetric pin on the shaft. If a common straight drive pin were used, it would be possible to mount the timing screw 180 degrees out of time. The asymmetric pin assures that it can be mounted in only one way.

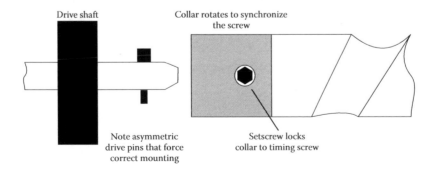

Drive shaft

Collar rotates to synchronize
the screw

Note asymmetric
drive pins that force
correct mounting

Setscrew locks
collar to timing screw

The following figure shows a starwheel with an adjustable hub. The hub is driven by two pins on the machine drive hub. Timing is accomplished, on first use, by loosening the bolts holding the hub to the star. Once timed, it should never be adjusted again.

Note that the drive pins on the machine hub are asymmetrically mounted. This prevents incorrect mounting of the starwheel.

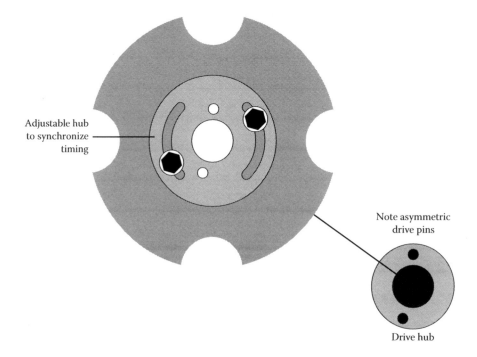

Adjustable hub to synchronize timing

Note asymmetric drive pins

Drive hub

Building the timing adjustment into the part rather than into the machine eliminates the time spent on synchronization as well as the time lost when it is not done properly. Perhaps the biggest benefit is that it simplifies the tasks enough that operators can now do them. This eliminated the time spent waiting for a mechanic. Mechanics are always in short supply and have more work than they can handle. This frees them up to use their time more productively.

Movable Motor Mounts

One company had a bottle feeder that had changeable paddle motor assemblies in the centrifugal feeder. They were about 8 feet off the floor, which necessitated working on top of a ladder. As they weighed about 40 pounds each, this created a safety as well as an ergonomic hazard. Due to the way they were mounted, changing them required tools, which meant that only a mechanic could do the task.

After evaluating the issue, different mounting brackets were designed and fabricated. One motor was mounted on a hinged bracket. This allowed it to be rotated out of the bowl. The other motor was mounted on slide bearings. A hand lever was loosened, the assembly slid forward into the bowl, and the hand lever retightened. A selector switch was added to connect the appropriate motor to the machine control panel.

It would have been possible to add air cylinders to move the assemblies into and out of position. Had this been done, all need to climb the ladder would have been eliminated. It was decided not to do this partly on the grounds of simplicity. A secondary reason was that they wanted an operator to visually inspect the bowl to make sure that there were no bottles from the previous run still present.

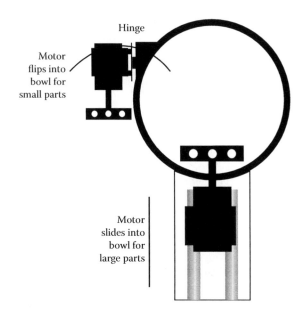

Sprockets and Gears

In some cases, machines will be mechanically linked but their relative speeds will need to be varied. One such case was a pouch-forming machine that was linked to a cartoner. Two product sizes were run. One required 3 pouches per carton, the other 2 pouches per carton. The machines were linked via sprockets and roller chain. During changeover, a mechanic was required to break the chain at the master link to remove it, remove 2 sprockets, mount 2 new sprockets, verify the relative timing of the machine when replacing the chain, and finally refasten the chain master link. This took a lot longer to do than to describe.

The drive system was modified so that 4 sprockets are always mounted on their shafts. The chain is longer than needed, with the excess slack taken up by spring-loaded tensioning sprockets. Changeover can now be done by an operator. After pushing the tensioning sprockets back, there is enough slack that the chain can be removed from one sprocket pair.

Timing marks inscribed on the sprockets are aligned with pointers affixed to the machine to assure that the 2 machines are properly synchronized. The chain is then placed over the sprocket pair to be run and the tensioner is released to take up chain slack.

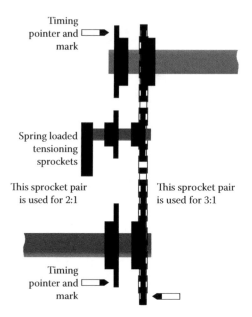

Gear-set timing can be difficult to reestablish when they are disassembled for cleaning or changeover. Alignment marks are critical to save time in reassembly as well as to avoid incorrect alignment, which can cause machine damage. This can be as simple as a center punching of a dot on a tooth as well as the corresponding land.

If there are several possible settings, they can be marked by stamping numbers into the gears at the appropriate point.

Conclusion

Albert Einstein reputedly said "Imagination is more important than knowledge." This statement is worth bearing in mind when working with LCO. We all share the natural human tendency to shy away from ideas when we cannot see for ourselves how to implement them.

This is a mistake. Coming up with the initial idea is often the hardest as well as the most valuable part of the creative process. An operator may come up with an idea for an improvement to a machine. They know what the desired end result is, but may have no idea how to achieve it. Perhaps it will require reprogramming the PLC. Perhaps it will require modifying mechanical parts to the machine. Whatever it is, they do not have and are often not expected to have the technical knowledge to carry it out. It is the idea that is critical. The technical development and implementation is a job for specialists such as engineers, programmers, and mechanics. All organizations need to drive out the fear that makes cowards of us all when it comes to suggesting changes and improvements.

A few years ago Adidas decided to make a running shoe that would automatically adapt to the conditions of the running surface. The shoe, called the Adidas 1, contained a sensor to determine whether the runner was on a hard, soft, or in-between surface, control buttons to set individual preferences, a servo motor to adjust heel firmness, a battery, and a microprocessor controller to manage all of this automatically. This entire package had to go into each shoe without significantly increasing the weight or degrading style and comfort. Finally, it had to do all this at some reasonable cost. It is unlikely that there is any single person anywhere who has the technical knowledge to do all this. It probably required half a dozen different kinds of engineers and designers to create. Initially, though, the idea came from the mind of a single individual. Whether that person knew how to design any of the components is immaterial. It might even be better that they have no technical knowledge because if they did, they might have decided that it was impossible to do and discarded the idea.

Changeover is like that. Coming up with the idea of what is needed or wanted is often the hardest part of the process. All managers, supervisors, engineers, and others need to encourage everyone in the plant not to be shy about their ideas. Not all will be good. All must be evaluated and the gems identified. In all cases, the person with the idea must be recognized. Failure to do so with an impractical improvement idea may well result in the next home run idea dying stillborn.

Whatever the mind of man can conceive and believe, it can achieve.

—Napoleon Hill

Chapter 5

Simplify

It is difficult to imagine a task or process that cannot be simplified. Everyone on the production floor must constantly ask themselves, "How can I simplify and improve this?" There must then be a system in place whereby these ideas will always be considered respectfully, and when they have merit, implemented. All employees must be encouraged to recommend improvements no matter how farfetched they might seem. They must not stop thinking just because it's 5 o'clock, either. Chances are that they will see, almost every day, ideas that are applicable on the production floor. They must keep their eyes open and bring these ideas to work the next day.

Smooth Surfaces for Cleaning

Consider the common kitchen stovetop. Some readers may have or remember older style ranges with individual burners or heating elements. As the sauce would boil over, it would run down into the burners and thence into the stove. Cleaning required disassembly and lots of elbow grease. Even then it still seldom seemed really clean. Modern ranges have a smooth, non-stick, ceramic top. Spills are easily wiped up, and with a minimal amount of cleaning, the five-year old surface looks as good as it did when new.

Production machinery and production spaces must be designed with ease of cleaning in mind. One factor that influences ease of cleaning is smoothness. Cracks and crevasses as well as sharp inside corners should be avoided. Where joints are unavoidable, if the panels do not need to be removed frequently, filling them with silicone caulking compound helps to prevent dirt infiltration. Rough surfaces can be painted with heavy "crack filler" paints to smooth them out.

When fasteners interrupt an otherwise smooth surface, they slow cleaning. In the case of a stud with a nut, the exposed threads will catch grime. If cleaned well, they will consume additional time. An acorn nut can often be used to cover the exposed threads. A bolt with hex head can be difficult to clean around due to the corners. A button-head bolt with a female hex may be a viable, and smoother, alternative. Fill the hex socket with silicone if it is not regularly removed. Best of all is a countersunk flathead screw or bolt. This leaves the surface completely flush for ease of cleaning.

Washroom

In the food, pharmaceutical, and other industries where parts must be washed during changeover, there are usually many opportunities to simplify the washing process.

Wash sinks are pretty mundane items that most people do not pay enough attention to. The sinks found in industrial washrooms are often standard sinks sold by kitchen supply houses. While this may save a few purchasing dollars, it will likely end up costing more in the long run from daily inefficiency. Perhaps the most important point is that the sink must be large enough for the parts that are to be washed. Skimping on the size can slow down washing. Construction should normally be of stainless steel for easy cleanability and resistance to staining. All corners should be coved or radiused rather than square to eliminate places for dirt to accumulate. Adequate drain shelves or tables, generally on both sides of the sink, must be provided for use as working areas. It may be useful to provide a shelf in the front of the sink as well. Sinks should normally be double, though this may depend on the specific cleaning process to be used. Double sinks allow one sink to be filled with detergent solution and the other to be used for rinsing.

Water connections should be oversized to speed filling the sink. If standard faucets are used, it may take 5 minutes or more to fill a 20-gallon sink. Be generous and use as large a faucet as possible, perhaps 1 or even 1½ inches to speed filling. An aerator or splash guard can help eliminate splashing when filling at high rates. In addition to the standard faucets for filling the sink, a large spray nozzle should be incorporated. These are commercially available and are usually suspended over the sink. These can be very useful for both pre- and post-rinsing the parts being washed. Again, don't skimp. Get one that will direct a high-pressure, high-flow stream on the parts being washed. It is better to be oversized and throttle it with a valve in the supply line than undersized and wanting more flow.

In many washrooms, the amount of water to be mixed with the detergent must be measured. An easy way to do this is to mark the sink with appropriate gradations. This should be a single line, if possible, to minimize the possibility of confusion on the part of the operator. An automated filling system that uses a flow meter to measure out, say, 20 gallons then shut off may be a useful option. This allows the operator to perform other tasks while the sink is filling. Drains should be oversized as well so that the used water can be emptied quickly.

Some washrooms have a single sink. This may be adequate for washing the parts during changeover. Even so, a single sink can be a choke point. If, while the sink is in use, someone else needs to wash something, perhaps a utensil that has fallen on the floor, sharing one sink can disrupt the washing process. It is a good idea to have at least one additional sink in the washroom for such casual usage.

Larger parts may not fit in a sink. These parts may be washed while on the floor or on carts. If no better provision is made, this results in a lot of water on the floor, causing a potential slip-and-fall hazard. The water also gets tracked about, either in the room or outside of it. One alternative is a washing grate or booth. This consists of a grate on the floor large enough for the parts to be loaded onto. The grate is mounted either in the floor or on a low box to collect the wash water. Side panels with grates can help control splashing. Appropriate pressure hoses and nozzles should be permanently mounted. Ideally, they should be ceiling mounted, on hose reels if appropriate. This will prevent the hoses from lying on the floor getting dirty and presenting a tripping hazard.

The proportion of detergent to water is critical to cleanliness. Too little and the parts will not get clean. Too much and residue from the detergent can remain on the parts even after thorough rinsing. In some cases, too strong a solution can cause discomfort and even health problems for the operators working with it. Detergents must always be mixed properly for best results. Some washrooms measure the detergent by volume using a measuring cup or graduate, others measure by weight, and either is acceptable. What should not be acceptable is having the operator measure the detergent each time they need to wash. Some suppliers offer detergents in single use packages. This takes the task and the potential error of measuring out of the operator's hands, as well as the lost time. If suitable detergents and packages are not available, the plant can make their own. Use an operator to prepare bags or bottles of detergent in advance when they have some slack moments.

Most washing will require tools to achieve cleanliness. Sponges or rags other than single use disposables should be avoided. When damp, they make excellent growth media for bacteria and mold, which can then be transferred to subsequent parts. They can also trap dirt that may not be removed by simple rinsing. It is generally preferable to use brushes. Brushes should be good quality, generally with plastic handles and synthetic bristles to maximize cleanliness. In some cases metal brushes may be required. As the saying goes, "One size fits all never does." Do not try to find a universal brush. It is far more likely that multiple brush sizes will be required for different parts and different cleaning requirements. It may be a good idea to get a cleaning supply catalog and order one of every brush size and style. Give them all to the operators and let them decide what works best, and then standardize. The brushes that don't make the cut in the washroom can always find a use elsewhere. If not, think of the extra brushes as inexpensive research toward finding the best tools.

Most of the time brushes are held by the operator and rubbed against the parts. Especially with smaller parts, it may be more useful to use fixed brushes and rub the parts against them. Restaurant supply houses sell glasswashing brushes that attach to the bottom of the sink with suction cups. These are commonly used in bars and restaurants with the glasses passed over them for cleaning. These brushes may be static or powered. If there are many small parts, a brush that can clip to the side of the sink, either above or below the surface of the water, may be useful.

Small wire-mesh dipping boxes are a useful tool for small parts such as nuts and bolts, which can be placed in the box and left soaking in a corner of the sink while the rest of the parts are washed.

Drying

Parts should be thoroughly dried after washing. The use of compressed air is a universally popular, but not terribly effective method for this. One problem is that the commonly used nozzles are designed more for blowing off particulates than water. In a few cases, drying may be attempted using a compressed air hose without a nozzle. This is not only ineffective, it is a safety hazard. A better tool for drying is an air knife. These use highly directed air to blow water away. Air knives may be handheld and used as wands. In some cases, where there are many similar parts to be dried, it may make sense to build a small device to semiautomate the drying. Pharmaceutical tablet fillers use a large number of long thin slats. One alternative for drying is to mount a small portal with one or more air knives. Rather than passing the air knife over the slat, the slat is passed through the portal and over the air knife, simplifying handling and manipulation of the slat.

Brushes can be particularly hard to dry. In one plant a mechanic came up with a device to speed drying of rotary roll brushes. A Lexan box was fabricated with 3 cam followers at each end, 2 in the bottom and one in the top. The brush shaft is laid on the 2 lower followers and held in place by the third when the lid is latched closed. A small motor mounted on the box is used to rotate the brush at a high RPM, flinging the bulk of the water out. A further refinement would be to add an air knife in the box to speed drying.

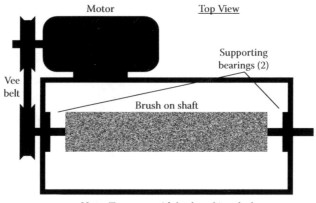

Note: Top cover with latch and interlock
required for safety. Not shown for illustration.

It may be possible to automate parts washing and, if so, this should be considered. Specialized and general-purpose washers are available but are often expensive. Cold glue labelers can be messy and time consuming to clean after use. In one plant, the mechanics bought a standard, consumer-grade, dishwasher and installed it by the labeler under a worktable. This allowed parts to be placed in the washer directly on removal, eliminating transportation to the washroom. The author's first reaction on seeing this was that it was not industrial grade and will fail within a year or two. While this is true, the dishwasher only cost about $400. When it fails it is easily replaced out of the maintenance budget. A comparable industrial-grade washer would have cost $20,000 or more. A plant would need to go through a lot of dishwashers (50 or so) before the consumer model cost more than the industrial grade. Once installed, the optimal arrangement of the parts was determined and photographed. This picture was laminated and prominently posted to help assure proper loading.

In all cases the washroom must be treated as a production work center. A 5S program must be instituted to assure that it is well organized. The teammates working in the washroom must understand their tasks' and how best to carry them out. Supervision must be in place to assure that they do. If labor loading permits, it can be a good idea to have teammates permanently assigned. This allows them to develop expertise and a rhythm in their work that may be difficult to achieve when washing is done ad hoc by operators from the line.

The washroom must be designed so that parts flow through in a consistent manner. Failure to do so can result in parts that have not been cleaned being mixed in with parts that don't need to be cleaned.

Part washing usually takes place away from the production floor. If redundant parts are available, it can become an external operation (see the following chapter) and the actual time spent in the washroom becomes much less critical. It always needs to be done as efficiently as possible. If it is done externally, since it is no longer delaying production, more care can be taken to assure that it is done effectively.

With few exceptions, cleaning on the production floor cannot be externalized. The focus must be on simplifying and improving to allow it to be done faster while maintaining quality. In many cases, the major opportunity for improvement will lie in better organization. This means studying the cleaning requirements and developing the optimal procedure for carrying them out. Once developed, these need to be thoroughly documented, the teammates trained, and supervisors need to assure that it is performed the same way every time.

Production floor design can have a big impact on cleaning by making it easier and in some cases reducing the amount required. One operation with a dusty product took place in a room approximately 100 feet long by 30 feet wide. All of the dust was generated in one small area by a single machine. Because of the room's open design, the dust wound up everywhere. This required cleaning of the entire room. A separate room, or even a sealed machine cabin, maintained under a slightly negative pressure, would confine the dust to a small area.

The lighting in this area was provided by standard fluorescent fixtures. These were not sealed and dust collected in them. This required two operators to clean them periodically, one on a tall ladder and a second to steady the ladder. Each of 45 fixtures was opened, wiped down, and reclosed. This was very time consuming. Worse, it was dangerous to the operator on the ladder as well as dangerous to the room if a tube had broken. These fixtures are not suitable for this type of area and should be replaced by sealed, dust-tight fixtures to eliminate the need for cleaning.

Ventilation air diffusers and returns also need to be designed for ease of cleaning. The typical diffuser is designed to optimize air circulation. In order to do this, it usually has a fairly complex geometry. Cleaning must usually be accomplished by hand at the top of a ladder. There are diffusers designed for cleanability with flat face panels. These should always be used because they are less likely to need daily cleaning, and when cleaning is required, it may be possible to accomplish it from floor level without the need for ladders.

Production rooms, whatever the product, should always be constructed with easy-to-clean surface finishes. Epoxy paints and floor finishes, beaded aluminum wall panels, plastic laminates, and other finishes can all help by being less susceptible to becoming dirty as well as being easier to clean when they do.

Production machines must also be designed for ease of cleaning. First, it must be easy to remove the parts remaining from the previous run. Removing small parts from a vibratory bowl feeder often means scooping them out by hand or even removing them one by one. An easier way is to incorporate a quick dump door. This is a trapdoor in the bowl that can be opened to allow parts to discharge into a secondary track leading to a box. These may be incorporated by the bowl builder or may be added later at the plant.

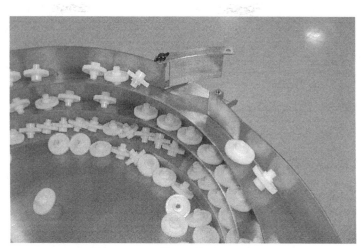

Photo courtesy of Service Engineering Inc.

Another alternative is to put a gate in the track or conveyor from the feeder to accomplish the same thing. Modify the conveyor rail to include a short, 6- to 10-inch section of movable or removable rail in a convenient location. At the end of the run, the rail is repositioned to run the containers off the conveyor and into a box. This saves the effort of removing the remaining containers manually from the hopper. If this is impossible, an alternative may be a door in the side of the hopper for easier access.

Photo courtesy of Service Engineering Inc.

Machine safety guarding must be designed to restrict access to the machine while in operation. This restriction should not extend to access during cleaning. Guarding must open fully or be easily removable so that it is not in the way. Doors used with guarding systems can be a real pain in the neck (*literally*) by not staying open and getting in the way generally. Some newer machines hinge the doors so that they open up, rather than out. This can be accomplished by mounting the door hinges at the top of the door or designing the access panels to slide up. Air springs hold them up and out of the way. It may be possible to modify existing machine guarding in this way. If not, another possible alternative is to use hinges that allow doors, once open, to be lifted off and set aside. While this improves access, it also raises an issue of what to do with the doors after removal. If doors are to be removed, a rack or cart must be provided to hold them.

Light curtains are specially designed arrays of photoeyes that may sometimes be used to eliminate guarding altogether. They form a continuous curtain of invisible light which, if interrupted, stops the machine. They work well when properly designed and used. One objection to them, especially if there is much foot traffic in the area, is that it can be easy to inadvertently stick an elbow into the curtain stopping the process. Where used, they should be mounted in such a way as to minimize this.

Note: Prior to making any modifications to any guarding or other safety systems, consult a knowledgeable safety engineer.

Guarding can obscure machine operation, especially in dirty processes. One trick for better visibility is to use either expanded metal or wire mesh instead of Lexan. This will not contain any contamination, but the mesh will not obscure vision. Painting the mesh a flat black will help with visibility. Lighter colors tend to focus the eye on the mesh rather than the operation behind it. For dusty operations when using polycarbonate panels, a light amber color may provide better visibility. The dust still collects, but it does not show up as much due to the color.

Some machine tools provide spinning glass panels in the guard panels. The cutting fluids that splash on the spinning disk are thrown off by centrifugal force, which keeps them mostly clean.

Conveyors

One description of a conveyor is "a box designed to collect gunk" and this is often accurate. Some plants will disassemble conveyors daily to clean inside. The typical conveyor is neither open (for ease of cleaning) nor sealed (which hides the need for cleaning). There are several conveyor designs that can help promote cleanliness.

Sanitary conveyors are common in the pharmaceutical industry but little known elsewhere. The typical sanitary conveyor consists of a U-shaped cross-section frame, mounted open side down. A pair of rails is mounted on standoffs on the top, closed surface so that they are about one inch above it. Any contamination that may fall onto the conveyor falls on the top surface under the chain where it is easily wiped up. Open frame conveyor design takes this a step further and opens the frame almost completely. This does not prevent debris from falling into the conveyor frame but does make it very easy to clean without disassembly. Conveyors in beverage plants are often constructed from four angles held in position by spacers. This totally open design allows the conveyor to be hosed off at any time with no need for disassembly. Some plants have systems to spray the conveyor continuously both for lubrication and cleaning.

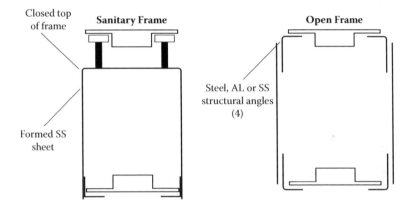

Stainless steel construction is required in some industries for reasons of cleanliness. Even when not required, it is a good practice to specify stainless steel or anodized aluminum for all new conveyors. They cost a bit more than painted steel. This additional initial cost is offset by reduced maintenance such as painting, better durability as they will not corrode, and better overall appearance as they will not chip or stain.

Several companies make continuous conveyor cleaning systems in two basic styles. One system is passive and passes the conveyor chain or belt though a trough of water, with or without detergent. This provides a continuous rinsing of the chain and may be sufficient in many cases. If not, for more stubborn contamination, an active conveyor cleaning system may be used. This consists of a scrubbing belt or brushes against which the chain is passed. Water and detergent is sprayed on the chain as it passes the scrubber and the combination of solvent and mechanical action keep the chain clean.

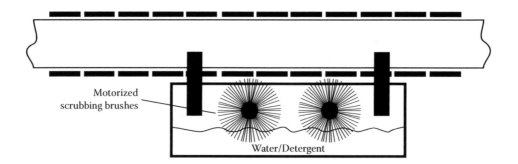

Motorized scrubbing brushes

Water/Detergent

Machine Cabinets

Another machine design feature that can facilitate cleaning is open frame construction. Closed frame machines can trap spills inside. Open or *balcony* frame construction allows any spillage to fall through the machine onto a catch tray underneath. Some bottle fillers are designed with the entire top of the machine concave. Spillages are collected toward the center and out a drain. This drain is connected to piping to either collect or dispose of the waste.

Cabinet bases on machines may not be as clean internally as they should be. One reason is the "out of sight, out of mind" syndrome. A related reason is that they may be dark inside and it is hard to see the dirt. Finally, access may be difficult, which further contributes to cleaning being ignored.

The first step in reducing cleaning effort is to eliminate or reduce the need for it. Machine cabinets should be sealed and slightly pressurized where possible to minimize infiltration of dust and debris. Door seals should be checked and replaced as necessary to assure tight sealing. Doors that are warped or latches that do not close tightly should be fixed or replaced. If the room is dusty and it is necessary to supply cooling air for motors or control panels, this should be supplied by ducting clean air to the cabinet from the ventilation system.

Operators may use flashlights to illuminate the cabinet during cleaning. One alternative to this is to mount small lamps inside the cabinet for illumination. These can be connected with a door switch so that when the cabinet is closed the light turns off automatically.

Clean-in-Place

Clean-in-place (CIP) systems allow for cleaning of machines, tanks, and piping without the need for removal to a wash area. These can be useful because they automate the cleaning, reducing the amount of labor required. If properly designed, they also assure that the cleaning is performed consistently. Eliminating assembly and disassembly also reduces wear and tear on the equipment and reassembly errors.

CIP systems can be a mixed benefit. On the one hand, they do improve cleaning. On the other hand, in most cases the CIP must be performed while the equipment is down. Some systems use multiple tanks or piping circuits, making it possible to CIP one while the other is in use. If changeover completion is delayed waiting for the CIP to finish, it loses its advantages in speeding changeover. If this is happening, it may be better to externalize the cleaning and use duplicate sets of parts. Externalization is discussed in depth in the following chapter.

Pigs

Pigs can speed cleaning of product piping systems. Pigs are rubber, foam, or plastic plugs that are introduced into the piping. They are tight fitting and, as they are propelled through the piping, they act as a squeegee, scraping the walls and pushing the product before them. Viscous products especially have a tendency to stick to pipe walls and significant amounts of product can be lost this way. The pig allows most of this product to be recovered while reducing the overall amount of cleaning necessary.

Pigs can be propelled through the system with compressed air or nitrogen prior to normal cleaning. Alternately, they can be propelled by the next product to be run. This would be acceptable in cases where slight mixing between products is acceptable. The first product continues to be run as the pig is pushing it out of the piping. When it arrives at the machine, a sensor detects the pig and stops production. The pig is removed and the next product can now be run. This cutover between products can be automated using sensors to sense the presence of the pig at the end of the line.

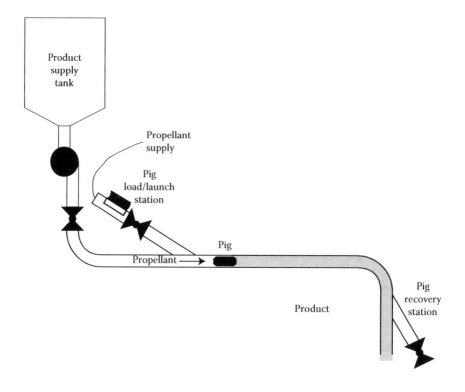

Tools

One issue in changeover is the availability of proper tools, with the emphasis on *proper.* The right tool should always be used for any job. There is even more reason to follow this rule in changeover as the few tools required are (or should be) known ahead of time. A set of tools required for changeover, and only those tools, should be provided and stored at the machine. Storage may be in a dedicated toolbox, but a better alternative is a rack or shadow board on the machine. These tools should be identified as belonging to that machine and should never be used for anything else.

Power tools can be an option for repetitive operations. A tablet press may have 80 or more stations, each requiring removal of a locking screw to change the die. These are often tightly fitting and require some effort for removal even beyond breaking them loose. A pneumatic ratchet wrench or screwdriver may be useful for removal of the bolts. The amount of torque applied when retightening the screws is critical. Too much can crack the die table, too little can cause the dies to loosen during operation. One solution is to use a torque wrench to break the bolts free, then use the power wrench to run them out and in. The torque wrench is then used for final retightening.

A change from a hand to a power tool for repetitive operations is not only a changeover time saver, it is also an ergonomic and safety improvement, reducing the risk of injury from repetitive twisting motions.

Tanks, reactors, autoclaves, and other pressure vessels may have their tops held by a number of clamps. A power wrench, such as a pneumatic ratchet and socket, can be a useful improvement here as well.

In another operation, a heavy machine had to be raised and lowered for different products. It was very heavy and required a big mechanic and a bigger wrench. Even then it was a lot of work. A ½-inch-drive pneumatic ratchet was mounted on the machine with brackets. The brackets were designed to support the wrench as well as prevent it from rotating. Raising and lowering can then be done by anyone in minimal time with minimal effort.

Many types of battery-operated tools are available, but pneumatic tools may be the better alternative in many instances. One reason is speed control. Restrictor plates or needle valves can be used to control air flow and thus the speed and force of the tool. Another reason is that they are less attractive and less likely to "walk away" from the machine. Finally, the air hose itself acts as a keeper to prevent walking off. If pneumatic tools are to be used, they should be dedicated to each machine. The air hose should be connected with threaded, not quick, connectors. The tool should be stored in or on the machine.

In a machine disassembly operation, there may be many bolts and nuts of varying sizes that require removal. A pneumatic ratchet or screwdriver can be useful, but it may require frequent changes of sockets to accommodate the different fasteners. The Gator-Grip™ socket may be useful in this situation. This is a 3/8-inch drive socket with a number of spring-loaded pins inside. The pins conform to the shape of any bolt head and even to irregular shapes. This allows a single tool to handle a wide range of fasteners without the need to change sockets.

Many plants will have mixtures of US (inch) and metric machines. This can cause confusion by requiring the use of different tools for each. Worse, it can result in the use, on purpose or unintentionally, of the wrong tool. Some inch-sized wrenches will almost fit a similarly sized metric fastener and vice versa. This can result in a temptation to use them interchangeably. However, when an inch wrench is used on a metric bolt or, worse, on a metric socket setscrew, it is likely to slip, damaging the fastener and potentially injuring the user. It also does not remove the fastener and often damages it permanently. Dedicated tools mounted on the machine will help. Color-coded tools can help mechanics quickly identify metric and inch tools. The company Metric-Blue sells blue Allen wrenches as well as blue fasteners, which are very helpful in identifying metric. Signs on the machine warning to use metric tools can also help. Some companies will mark blue circles around each metric bolt. This can be especially helpful when a single machine contains a mix of inch and metric fasteners.

No Tools

There are a number of ways to improve tool usage as discussed above. The better approach is to eliminate all tool usage wherever possible. There may be some exceptions to this. Tasks that the operator should be prevented from attempting may find the requirement for tools helpful. In other cases there may be safety issues that prevent the elimination of tools. In general though, while tools can be very useful, in changeover they can be a source of a number of problems:

- Inch tools may be used on metric fasteners causing permanent damage.
- If the proper tool is not available (and sometimes even when it is), improper tools such as pliers may be used. This will cause permanent damage.

- If pliers are used and they slip, they can generate metal particles. These can potentially find their way into delicate machine parts, causing premature failure. Product contamination is another possibility.
- Tools are a consumable wear item but may not be recognized as such. This can result in tools such as hex wrenches and screwdrivers being used beyond their useful life. When this happens they can damage the fasteners, making them permanently difficult to remove and perhaps making it impossible to use the proper tool.
- Time is wasted walking back and forth between machine and toolbox getting tools.

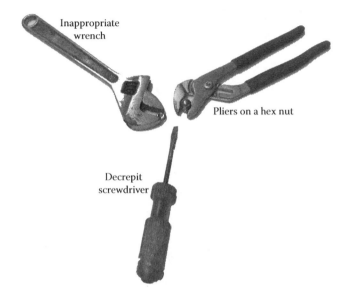

Inappropriate
wrench

Pliers on a hex nut

Decrepit
screwdriver

Some plants, by policy or by labor contract, do not permit operators to use tools. Operators may be permitted to perform some simple changeover tasks but only if no tools are required. Substitute a hand knob for a bolt and there may no longer be a need for a mechanic to make that adjustment. The mechanic is not idled. Allowing operators to perform rote tasks frees the mechanic to do more valuable tasks such as machine repair and maintenance and machine improvements.

One of the first steps in any lean changeover program is to identify all the tools that are required in the changeover, where they are used, and why. Once this is done, each instance should be analyzed to determine if there is a tool-less alternative available. If so, it should be evaluated and implemented.

Ideally, all routine changeover should be tool-less. Additionally, all non-changeover adjustments and fasteners should generally require a tool. This provides a visual aid to the person performing the changeover. When a tool is required, it reminds them to think twice about whether they should be making that adjustment or removing that part.

Hand Knobs and Levers

There are many devices available to eliminate tools. The first ones that come to most minds are hand levers and hand knobs. For the purpose of this book, hand knobs will be defined as symmetrical and hand levers will be defined as asymmetrical. There may be some exceptions, but in general it is best to use hand knobs where the fastener must be removed and hand levers where the fastener only requires loosening. One reason for this is ergonomic. Asymmetric hand levers can be difficult to unscrew and even more difficult to restart without cross threading. Hand knobs, because they are symmetrical, avoid this issue.

The second reason is mnemonic. If this policy can be applied consistently and broadly and if the operators and technicians are trained on it, they will have a pretty good idea from the fastener style what to do with it. "Round knob-remove/Lever-loosen."

Hand knobs come in a variety of shapes—round, triangular, star, and others. Round knobs are good when normal tightening is required under clean and dry conditions. If the hand knob is likely to be wet or slippery or the operator is likely to be wearing gloves, a knob with a star or other nonround shape will provide a better grip. T-handles, with some exceptions, should be avoided. They combine the worst features of both knobs and levers without much redeeming value.

Levers also come in a variety of styles and sizes. One commonly used style incorporates a spring-loaded spline in the lever. This allows it to be lifted up and repositioned axially to avoid interference with machine components. Some less-expensive hand levers are made with plastic splines. These may be fine in some light duty, nonindustrial, applications, but they tend to strip easily when used daily or when firm tightening is required. It is recommended that hand levers with stainless steel internal components always be used. Handles may be metal or plastic and both seem to hold up well in an industrial environment.

Photo courtesy of JW Winco

Occasionally the fastener will be deep inside a machine. In these cases, hand knobs/levers with shoulders can be used that extend the handle out to an accessible area. Note the nonthreaded lead-in. This is a good feature on all threaded fasteners as it makes cross threading almost impossible.

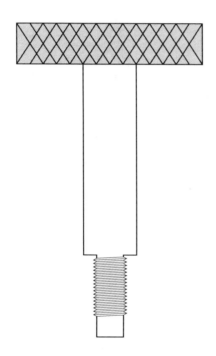

Quick-Acting Knobs and Levers

Several designs of quick-acting hand knobs and levers are available. One is a standard knob drilled out at a slight angle. This angled bore is sufficient to slide over the outer thread diameter. In use, the knob is loosened about ¼ turn and allowed to tilt slightly. It can then be slid over the threads and removed or repositioned. When retightened, it straightens, engaging the main thread, and a ¼ turn locks it in place.

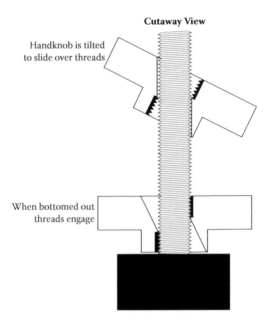

Another style uses a threaded half-nut, spring loaded and mounted inside the knob or lever. To remove or reposition, the knob/lever is turned ¼ to ½ turn to release pressure and a button is depressed which disengages the half-nut from the threads. The knob/lever can then be slid over the bolt threads and removed. When replaced, the procedure is reversed.

Still another style is a nut that mounts the threads on pivots with a locking collar. The nut is loosened and the locking collar lifted. This releases the threads, allowing the nut to be removed.

Open, threads disengaged   Closed, threads engaged
Photo courtesy of JW Winco

When using either levers or knobs, it is recommended that they be purchased in bright orange (often called "safety orange"). This provides greater
visibility for the operator, jumping out at them.

Cam action clamps can be used where tight, friction clamping is required.
Cam clamps use an eccentric cam mounted on a lever to apply pressure to
a gripping block. Cams are usually not positive locking and can slip under
pressure. In some applications this may be desirable so that in the event of
a jam or the like the cam slips before the machine breaks. In other applications this may not be acceptable.

Spring-Lock Collar

Spring-locking collars use internal spring-loaded balls to provide clamping
against a shaft. The balls only lock in one direction, which allows the collar to be slid on, locking in place automatically with a strong friction fit. To
remove the collar, the ring is grasped and pulled up, releasing the spring
tension on the balls and allowing the collar to slide off. Some models allow
the locking ring, when raised, to be locked in the open position. This can
be handy when there are multiple collars in use. Other models have an
external thread and/or mounting flange. This allows them to be permanently
mounted on the part.

Pins

A number of different styles of pins can be used to mount machine components.

Hitch pins are bent out of spring wire. The straight portion of the pin holds the component in place; the bent section goes against the component to hold the pin in place. They can be used in many lighter applications where it is not necessary to hold the parts tightly together.

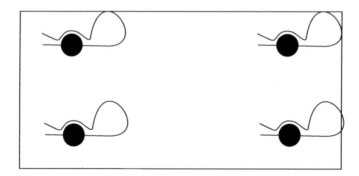

Ball pins are straight rods of appropriate size with a spring-loaded ball extending to the side near the end. The ball provides some resistance to the pin vibrating out of its hole. A drawback to this type of pin is that it relies on spring pressure to force the ball out. It does not provide positive locking.

The locking ball pin is an improvement on the basic straight pin in that it positively locks the ball in place. A plunger on the end of the pin must be depressed to release the ball for removal of the pin. Note the keeper wire on the pin in the illustration. In one sense, this may be the most important component. This is about the only thing that prevents the pin from being misplaced after removal.

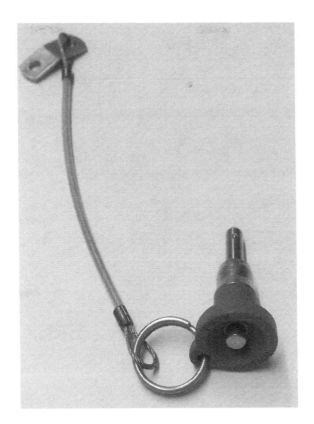

One application for these pins is mounting sheet metal panels. If two thin panels are pinned together, the pin may cock, letting the panels get slightly out of position. To avoid this, a strip of sheet metal can be welded to one of the panels. This allows the other panel to be slid in between and pinned so it cannot move.

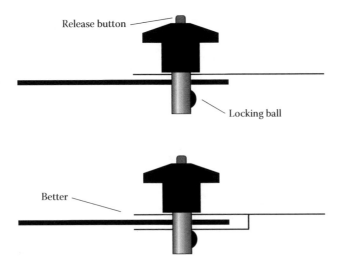

Still another type of pin is the plunger pin. This pin is permanently mounted to the machine or part, usually by threading into a hole. The tip of the pin is extended by spring pressure to hold the part in place. Some styles provide for locking in the retracted position, usually by rotating the pin ½ turn after it has been pulled back. Others are always extended except when being manually held open. A nonlocking pin has the advantage that it will not be inadvertently left in the retracted position.

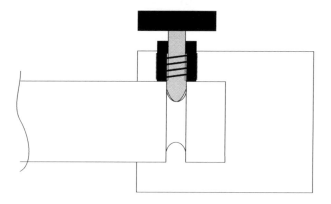

Snaps

Ingenuity is always a key factor. In this cartoner infeed "bucket" the width is varied for different parts by means of a spacer block. Rather than bolt the spacer in place, it is held by two snap fasteners. This allows them to be easily popped in and out of place. As this machine has a couple dozen buckets, this saves considerable time and ergonomic risk compared to bolted spacers.

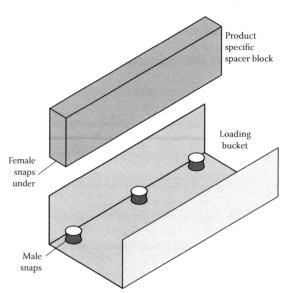

Magnetic Locking

Magnets may be used for mounting machine parts. In the illustration, the pusher is a changepart, changed for various size products. It is primarily held in place by the interlocking geometry of the pieces. The normal direction of force in operation forces the pusher back, locking it in place. Magnets are used to hold it there when there is no external force preventing it from moving out of position because of vibration. In this type of application the magnet provides minimal force, with the bulk being provided by the interlocking nature of the pieces.

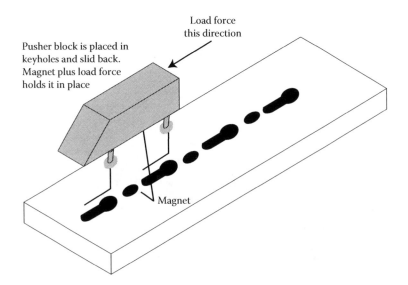

Magnets can also be used to provide the major holding force. Surface grinders and mills commonly use magnets as the principal clamping mechanism. These magnets can be electromagnets, which clamp on application of electrical power. Alternately, they can be powerful permanent magnets. These are always magnetized and the clamped part is released by moving the magnet away from it.

Vacuum Mounting

Vacuum can be a powerful tool for mounting components. This picture shows a vibratory feeder bowl using the Chase-Lock™ mounting system. Normally the feeder bowl is mounted to its base drive using bolts or bolted clamps. In this system, the bowl is permanently mounted to a slightly concave baseplate. The feeder base drive has a corresponding plate mounted to it. An "O" ring provides an airtight seal between the two plates. A vacuum port in the lower plate allows vacuum generated by a venturi generator to pull the plates together. A pressure switch turns the generator on and off as needed and also acts as a safety device to stop the machine if the vacuum falls below a certain level. On a 15-inch-diameter mounting a vacuum of 25 inches of mercury provides more than 1000 pounds of clamping force. This plate also has guides to locate the bowl plate. The bowl is mounted by setting it on top of the lower baseplate. This design can also be adapted for quick mounting of other components.

Photo courtesy of Chase-Logeman Corporation

Toolless Plunger Mount

Many will be familiar with the Craftsman Speed-Lock™ drill chuck ad that runs on TV. This chuck has a spring-loaded collar that is pulled back, allowing a matching drill bit or screwdriver to be inserted. The collar is released and the bit is held in place by internal balls and springs. This concept can also be adapted to facilitate changeover. The following picture shows the stopper insertion plunger on a Chase-Logeman stoppering machine, which was inspired by the Speed-Lock design. Previously, two wrenches, one for the locking nut and one for the chuck, were required. As the chuck is in an awkward position and the machines are commonly used in cleanrooms, this was a difficult part to change. Now it pops in and out with no tools.

Spring loaded collar slides up to release plunger

Photo courtesy of Chase-Logeman Corporation

Quarter-Turn Fasteners

Quarter-turn fasteners can be useful for cabinet panels that need to be removed regularly. They are available in a variety of grades, sizes, and styles to fit every need. They can be configured to use a standard or a special tool to minimize access. They are also available with thumbscrew or knob head for tool-less removal.

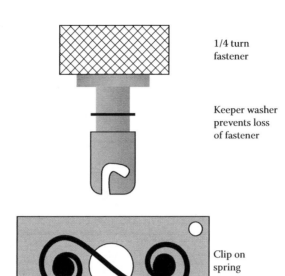

1/4 turn
fastener

Keeper washer
prevents loss
of fastener

Clip on
spring

Tool Changers

Robots are often expected to perform multiple functions on the plant floor. As they switch between functions, it may be necessary to change the end effectors (such as a gripper or welding head) for the different products and functions. This can involve unbolting the effector and disconnecting pneumatic and other hoses as well as control and power connections. Attaching and detaching all of these can be time consuming and invites the potential for error, especially when remounting.

End-of-arm tool changers consist of two components. One is permanently mounted on the robot arm with all electrical, sensor, and plumbing connections. The other is attached to the effector. A ball lock allows them to positively mate together. As they mate mechanically, the plumbing and wiring connections also mate. This eliminates any possible errors from manual connection. In some applications, no human interaction is required, with the device picking up the corresponding effector from a rack.

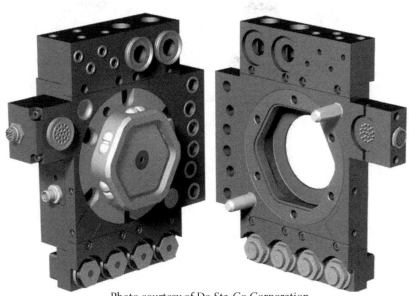

Photo courtesy of De-Sta-Co Corporation

Manual versions of the above are also available. These are similar in principle with only one connection required to be made. Once mated, a toggle lever locks them in place. Plumbing and electrical connections mate as the two halves are mated and a manual latch holds the two parts together.

Toggle Clamps

Toggle clamps are available in a wide variety of sizes and configurations. Toggle clamps can be powered or manual. They can be either single-acting, locking only in one position, or double-acting so that they can be locked in either the extended or retracted position. Toggle clamps consist of a holding arm and an actuator arm. The actuator arm forces the holding arm into position, then continues moving over center to positively latch into position. This provides a more positive latching than the cam clamp, which applies incremental pressure. This may be a disadvantage in some applications if the toggle clamp is used to hold parts of different sizes in position. Toggle clamps, once set up, have no adjustment or play. They are either fully latched or fully unlatched.

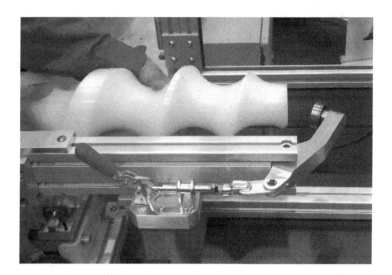

Toggle clamps with locking levers are also available and are recommended, especially for critical applications. They are similar to standard toggle clamps but have an additional locking lever. This lever engages the actuating lever when in the closed position, preventing it from being released inadvertently. To release the toggle clamp, the locking lever is first released, allowing the actuating lever to be released.

Pneumatically and hydraulically actuated toggle clamps are also available. These can provide greater holding force where required. They can also be automatically opened and closed if desired.

Conical Cylinder Locks

Conical cylinder locks consist of a ball end on a conical male piece. These are typically mounted on the removable component. The male section is inserted into a female section, which grips the piece with spring-loaded ball locks. This provides the dual benefit of very precise location due to the matching tapers as well as positive locking by the balls. The female sections use pneumatic pressure to open with the spring to lock, providing a fail-safe benefit. This assures that there is no failure if there is a loss of compressed air. The air can be connected permanently and controlled by a valve. For greater security, the lock can be fitted with a pneumatic quick connector with air connected only for unlocking. This prevents inadvertent actuation during machine operation.

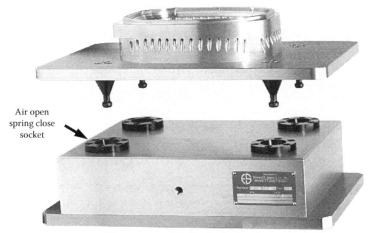

Air open spring close socket

Photo courtesy of Tooling Technology Inc.

Interrupted Threads

Intermittent or interrupted threads have long been used in many applications to allow quick assembly and disassembly of threaded parts. Artillery cannon breeches are one example. The threads of the two components, male and female, are each relieved in three sections. This allows the two parts to be quickly fitted together, and then a partial turn to tighten and lock.

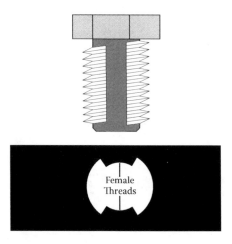

Female Threads

Slots and Keyholes

A key to simplification is to avoid the removal of any fasteners unless absolutely necessary. A part that mounts with a bolt through a hole means that the bolt must be removed to remove the part. Bolts that are removed are bolts that will be misplaced with time wasted looking for them. Worse, a bolt that is removed can fall into the machine, causing damage and even more serious time loss.

If the hole is near the edge, it is often possible to convert it to a slot. This allows the bolt to be loosened rather than removed and the part slid free. If the hole is too far from the edge, a slot may be impractical. In this case, rather than a slot, a keyhole (sometimes called a *pear-shaped* hole) locates a hole large enough to pass the bolt head next to the hole. As with the slot, the bolt is loosened and the part is slid to the side. It is then lifted off with the bolt head passing through a keyhole. The sketch that follows shows a machine subassembly. This subassembly needed to be changed for different product sizes and configurations. It weighed about 40 pounds and was in an awkward position. As built, it was mounted with 4 bolts, all of which needed to be completely removed. Removal and remounting required 2 people, one to hold it and one to unbolt it. Converting it to slots and keyholes allowed it to be held in place with 2 hand levers. After modification, one person could loosen the hand levers and lift the assembly up about ½ inch forward to clear the bolt heads out of the machine. This eliminated the use of tools, eliminated the use of a second person, and most importantly, reduced the ergonomic hazards.

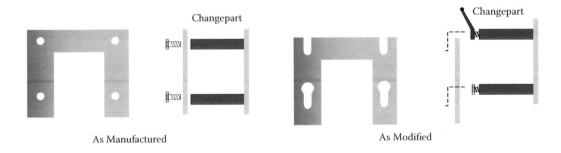

As Manufactured   As Modified

Note the position of the slots. The subassembly is held in place by the hand levers and studs, but if they should loosen, gravity will prevent a catastrophic failure.

Variations on this idea can be found in other applications. This figure shows a pressure tank cover where the round bolt holes have been reshaped into keyholes.

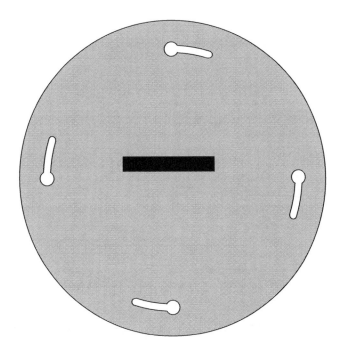

Another variation uses a slotted lid, or in the illustration, slotted pieces extending out from the lid, with hinged bolts. After the cover is mounted, the bolts are swung up and tightened.

C Washers

C washers can be used to accomplish a similar function where keyholes are not feasible. The part's mounting hole is oversized so that it can be mounted over the bolt head. A special heavy-duty C-shaped washer is slipped under the bolt head to allow tightening. A disadvantage of this type of mounting is that the bolt hole is too large and does not precisely position the part. This can be compensated for by adding locating pins or centering guides. Another disadvantage of the simple C washer is that it is another loose part that must be managed. An improved design adds a tab to the washer that allows it to be permanently mounted. The washer pivots out of the way once the bolt has been loosened.

Open for removal

Closed for use

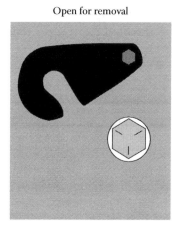
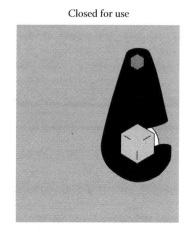

The illustration below shows an application of the C washer concept. In this case the part, a guide, is mounted on two rods, each of which has a groove machined in it. The guide is placed over the rods, the C washers swung in to engage the groove, and levers lock each rod in place with a cam action. This allows the guide to be changed in less than 30 seconds. Previously the guide had been mounted with a bolt threaded into each rod. Due to the location as well as the working conditions (typically a cleanroom with the mechanic fully gowned, hooded, and gloved), it was difficult to change and could take 10–15 minutes, longer if they fumbled the bolt. An additional issue, due to the location, is that the mechanic must wear sterile rubber gloves. If these get torn while they are fiddling with the guide and traditional bolts, it breaks sterility. Resanitizing may lose another hour or two.

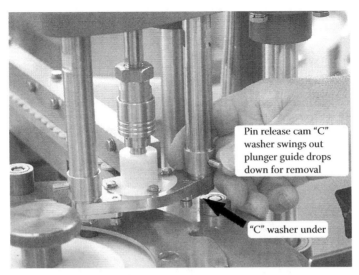

Pin release cam "C" washer swings out plunger guide drops down for removal

"C" washer under

Photo courtesy of Chase-Logeman Corporation

One-Touch Guide Rail Adjustment

Conveyor rails and other guides are usually not difficult to adjust but can be time consuming. There are several ways to simplify this.

Typically, chain-type conveyors, because of potential backlogs of product, will require more rigid rails and more closely spaced brackets (3–4 feet of linear spacing is typical). Vertical adjustment of the rails should be avoided. If taller products require raising the guide rails, consider double rails that can serve both sizes without vertical adjustment. Another alternative is to use wide plastic inserts on the rails to provide additional support. Guide rail brackets must be provided with hand levers rather than bolts for ease of adjustment. The exception to this is where the rail is not normally adjusted. In these cases, a setscrew is recommended to help prevent unnecessary movement of the rail.

Typically, conveyor rails are adjusted by a mechanic placing a bottle between the rails, adjusting them until they touch the bottle, then backing them off a bit. A problem can be that this *bit* will vary from person to person. One way to solve this it to wrap 2–4 layers of masking tape around the bottle to provide the desired outside diameter (OD). Now the rails can be adjusted until they touch the tape and the clearance will be appropriate for the normal, untaped, bottle.

Several manufacturers offer single-point conveyor adjustment systems. One such system consists of brackets with a rack and pinion enclosed in a plastic housing. As the pinion turns, it moves the rack, to which the rail is attached, in and out. The adjusting brackets are linked together by shafting. A right-angle drive on the shafting allows all brackets to be adjusted synchronously. A digital indicator or a set of stops (discussed further in Chapter 7) can be added to allow adjustment to a specific setpoint. A servo motor can be incorporated to allow for automated adjustment.

Photo courtesy of Septimatech Group Inc.

Another architecture for single-point rail adjustment mounts shafts above the conveyor with arms extending down. Guide rails are mounted to these arms and, as the shafts are rotated, the arms swing in and out. This is a very simple system, but because the arms move on a radius, it may be limited in range.

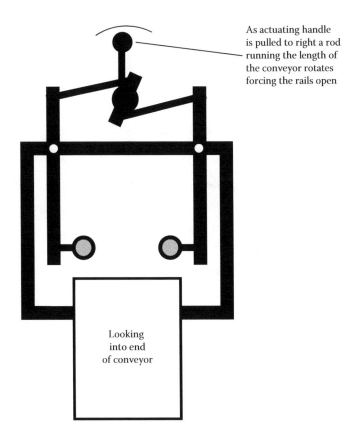

As actuating handle
is pulled to right a rod
running the length of
the conveyor rotates
forcing the rails open

Looking
into end
of conveyor

Rails on conveyor curves can be difficult to adjust because as they are adjusted in and out, their radius must change. In many cases, tight precision of guide rails is not needed and this is not a big issue. In other cases, such as with oval bottles that may shingle, jams may occur, making it an issue. There are several ways to handle this.

One method is replaceable curved sections sized for each container. These may be cut from plastic sheet and may be single or double high depending on the product height. Brackets are permanently mounted on the conveyor curve with pins over which the plates are placed. Often no fastening is used, relying on gravity to hold the plates in position. On high-speed lines or where excessive pressure against the plates may occur, it may be necessary to add toggle clamps or other devices to positively fix the plates.

If single-point adjustment rails are to be used, it is desirable that the adjustable portion extend around any curves. One way to do this is to use segmented rails that can adjust their radius as needed as they open and close.

Where products transfer side to side on overlapping conveyors, the guide rail adjustment has a similar issue to the curves. One way to resolve this is to build an offset into the conveyors so that the product never changes direction when transferring from one conveyor to another.

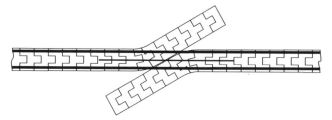

Straight line transfer, chain to chain

Single-Point Adjustment (Double Lead Screw)

Some machines use a pair of pneumatic cylinders to index the product into position. The downstream cylinder stops the group of products in position. At the end of the cycle, the upstream cylinder holds the succeeding product back until the first group has a chance to escape. Normally these cylinders need to be positioned individually, and it can sometimes require a bit of trial and error to get them right. A double lead screw (one half with right-hand and one half with left-hand thread) can be used to adjust both cylinders simultaneously while keeping them on a common center. The air cylinders are mounted on sliding brackets with the screw through them. As the screw is turned, they move toward or away from the center. A digital indicator on the screw allows it to be set to a precise numeric position.

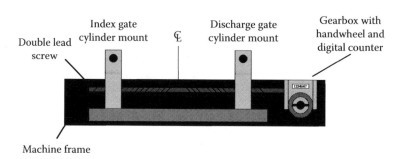

If there are more than two components to be evenly spaced, such as filling nozzles, the double lead screw can be combined with a pantograph.

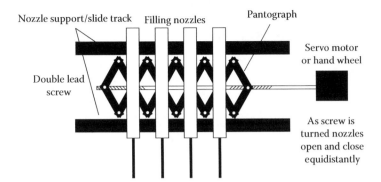

One of the concepts from Shingo's book on single-minute exchange of die (SMED) is that many bolts have excessive threads. The bolt must be long enough to go completely through the nut with one thread showing.

(labels in diagram: Nozzle support/slide track, Filling nozzles, Pantograph, Servo motor or hand wheel, Double lead screw, As screw is turned nozzles open and close equidistantly)

Captive Washers

If bolts with washers must be removed during changeover, the washer is an additional part that can be misplaced or perhaps accidentally dropped into the machine. If possible, reduce the probability of loss by using either flange nuts and bolts or nuts and bolts with captive washers.

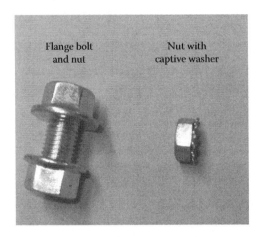

(labels in photo: Flange bolt and nut, Nut with captive washer)

One of the concepts from Shingo's book on single-minute exchange of die (SMED) is that many bolts have excessive threads. The bolt must be long enough to go completely through the nut with one thread showing. Any exposed threads have no function except to waste time running the nut down over them. Bolts should be properly sized so that no more than one thread is showing when properly tightened. Even better, any bolt that must be routinely removed should have the end relieved to the root diameter. This will greatly ease threading during reassembly. The relieved section should be 1–3 diameters long and relieved sufficiently to loosely slip into the female threads. Once in, it guides the threads, assuring easy mating with greatly reduced chance of cross threading. This not only saves time replacing the bolt, it reduces potential damage to the machine, which will be very time consuming to repair.

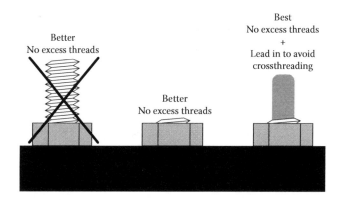

Unitary Lane Spacers

Lane dividers on case packers commonly use individual spacers to provide the correct separation between lanes. In some instances several dozen spacers are required. One way to reduce the number of loose parts is to make monoblock spacers. Instead of a series of individual "V" spacers, a single spacer can be machined from a block of plastic. Slots go over the lane guides and provide the correct separation. Multiple spacer sets can be machined from a single block if there are not too many different lane patterns. This allows all spacers to be permanently stored on the machine, reducing the time spent fetching the appropriate set for each changeover as well as reducing chances for damage.

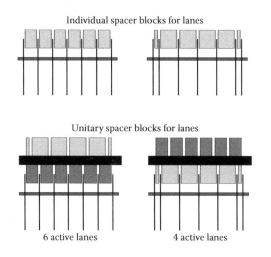

Asymmetric Parts

Poka-yoke is a Japanese phrase for mistake proofing. Parts that are removed and replaced or changed for changeover need to be mistake proofed. If there are two positions in which a part can be mounted, it is almost guaranteed that it will be mounted incorrectly from time to time. All machine components must be designed so that they can only be mounted in the correct position. This can usually be accomplished by making mounting holes asymmetrical. If the mounting holes already exist, a pin with corresponding hole can be added to prevent improper mounting.

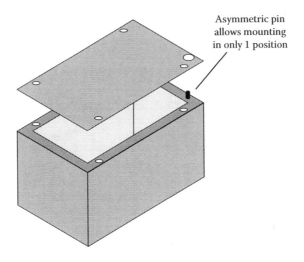

Asymmetric pin allows mounting in only 1 position

Scissor Jacks

When parts need to be raised or lowered during changeover it can sometimes be cumbersome. The part may be clumsy, making it hard for one person to hold in position while they tighten it. It may be heavy. It may simply, due to length or other factors, be difficult to align. One alternative is scissor jacks. These look like the jacks found in cars and are available from industrial supply houses in various sizes. Screw jacks may also be used in some applications. In the example, a 10-foot-long sheet-metal guide was held in place by 3 hand knobs and had to be aligned vertically to the product. Even with two people, this was time consuming and fiddly. A pair of jacks was mounted on brackets under the cover. They were connected by a rod that passed through the end of the cover and terminated in a crank. One person can loosen the knobs, crank the guide up or down into position, then retighten the knobs. The guide is perfectly aligned and adjustment takes less than a minute.

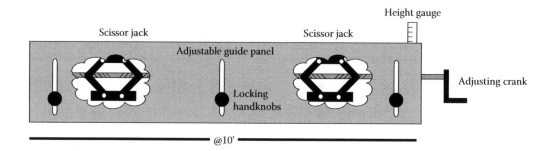

Changepart Identification

All machine parts that are changed during changeover (i.e., changeparts) must be properly and visibly identified. It must be readily apparent what machine, line, and product they are for. Color coding is one easy way to accomplish this. In the case of plastic parts, the parts themselves can be machined from colored plastic. If this is done, it becomes easy to spot an incorrect changepart.

It is not often possible to color code the entire part. As an alternative, color-coded dots can be used. A hole 1/16 to 1/8 inch deep is milled in the part and the bottom filled with a colored epoxy. Multiple and/or shaped holes can be used for additional information.

If neither of the methods above is practical, a plastic or metal tag can be engraved with the appropriate information either textually or pictographically to tell where the part is located. This tag should be permanently affixed to the part where it will be easily viewed both in storage and when the line is in production.

Bar and radio frequency identification (RFID) coding may also be considered for tracking changeparts. Bar codes are available in a variety of one-dimensional (1-D) and two-dimensional (2-D) formats. The 1-D barcodes have the advantage of simplicity for reading and printing. Although they can only encode a couple dozen numbers and letters, this is normally sufficient for identifying the parts. If more information needs to be encoded, 2-D codes are available that will encode hundreds and even thousands of alpha-numeric characters.

RFID is similar to a barcode in that it stores machine-readable data on each part. It differs in that the code is transmitted by radio waves, eliminating the need for individual scanning with a reader. Any RFID coded part, whether by itself or on a cart with many other parts, will automatically be scanned as it exits and reenters the storeroom.

Conclusion

There is almost always a simpler and easier way to perform any task. Simpler and easier also means faster as there will be less wasted time. One of the challenges of a lean changeover (LCO) program is to find the easier way.

Many solutions may be found while wandering around outside of work. One mechanic found inspiration in the thermoformed trays that a fast food restaurant uses to store parts for the milkshake machines. He found a company that could make thermoformed storage trays for a variety of small machine parts that he had trouble managing.

An operator found the industrial-grade wall-cleaning mops used in her plant uncomfortably heavy. She convinced her plant to change to a household mop with a light aluminum shaft and disposable pads. These do not last as long as the industrial-grade mops. On the other hand, they do not cost anywhere near as much, making them cost competitive even with more frequent replacement. The real benefit is that it makes the cleaner's jobs easier, resulting in better and faster work. The lighter weight also makes them ergonomically safer.

A quality inspector was bugged by always needing to look for a flashlight to inspect inside a machine cabinet. She wondered why it could not be like her refrigerator where the light went on when the door was open—a very simple solution and quickly implemented.

These are only a few of hundreds of examples that the author has seen firsthand. There are probably thousands more. The key is to turn people loose to find ways to do their jobs better.

Be lazy. Find the easier way.

Chapter 6

Externalize

Perhaps the greatest contribution of Shigeo Shingo's single-minute exchange of die (SMED) system was the concept of converting changeover tasks from internal to external, sometimes called *intrinsic* and *extrinsic*. An internal task is one that is performed while the line is stopped. Shingo teaches that tasks must be converted to external wherever possible.

The goal of lean changeover is to maximize the amount of time the line is producing good product at normal speed. By itself, externalization does not eliminate any tasks and may not save any labor hours. In some cases it may even increase the labor hours. An operator may be available to fetch parts during changeover if the machine is stopped. If the task of fetching parts is done prior to changeover, the operator will not be available to do it and an additional person may be required.

Externalization will still make economic sense due to the disparity between the labor and the cost of downtime. It is likely that the downtime will cost $10,000 or more per hour. In most parts of the United States, an operator will cost in the range of $15–30 per hour. A technician such as a mechanic might cost double that. At this differential it will take a lot of operator and mechanic labor hours to make up for the cost of even a single hour of machine downtime.

The need for every plant to know the costs of changeover downtime were discussed in Chapter 2 and cannot be over emphasized. Frequently, externalization will require the purchase of additional tooling, changeparts, specialized handling equipment, or even entire duplicate machines. If manufacturing requests authorization to spend $50,000 on additional parts to save 20 minutes per day, management has no way of evaluating whether that is a good or a bad use of capital. Management's default decision in a case like this is probably to do nothing and rightly so.

If the cost of downtime is known by management to be $10,000/hour, the calculation becomes straightforward. Twenty minutes/day is 4,800 minutes or 80 hours per year. That equals $800,000 in annual savings in return on an investment of $50,000. This is about a 4-month payback on investment. Most managers would think that well justified.

Some externalization opportunities are easy to see and free or nearly free to implement. Movement of materials prior to rather than during changeover is one example. Others may be costly and complex. Many will fall somewhere in between.

Externalization can occur prior to production or post production. Preparing materials in advance is one example of prior externalization. Performing parts cleaning or product reconciliation after line restart is an example of a postproduction externalization.

Material Handling

Part, component, and material movement from the warehouse can represent a significant opportunity for externalization. It is not uncommon to complete the changeover before fetching the materials required for the next production run. In some cases, the materials have not even been picked from the warehouse racks prior to this point. Assuming that everything goes well, there is a fairly obvious loss of production time from this practice. Assuming that everything goes well, every time, is probably not realistic. When the materials are requested there can be a number of delays as a result of other orders being picked, a dead battery on the forklift, warehouse operators on break, errors in documentation, or missing signatures.

A more serious problem can occur when the inventory control system does not match what is on the racks. This can be the result of errors in the system if it is not updated and checked rigorously. The system might show 5 pallets of parts when there are only 4. Or there might be 5 pallets, but one pallet is missing 2 or 3 cases that were removed for whatever reason. A related problem can occur when all 5 pallets are present but are not available for use. This might be because they are awaiting inspection approval, have been damaged in storage, or have expired.

In one plant they would normally complete the changeover prior to picking the material in the warehouse. Frequently, at least once or twice a week, the materials would not be found in the warehouse. In some cases, the materials were misplaced and eventually discovered after time was wasted on a search. In other cases they were not found and more time was wasted making a decision about what product they had materials for and could run. Only then were they able to perform a second changeover to run the parts that they had. No matter how efficient changeovers are, anytime that there needs to be 2 changeovers to run 1 product, far too much productive time is wasted.

All components for the succeeding production order must be picked while the previous order is still running. This will permit any problems to be discovered and corrected before they can affect production. They should be staged as close to the production area as space and regulatory considerations will permit. If they cannot be staged near the line, one alternative is a dedicated staging area in the warehouse.

One plant goes even further and picks 2 orders ahead. The raw materials warehouse has a section where components for the next run are staged. In this plant they have 7 lanes marked on the floor, each corresponding to a production line. The materials for the next order will always be on pallets at the discharge end of the lanes, near the warehouse door. All documentation is prepared and staged on top of the end pallet. Placing the materials on the floor on pallets permits them to be removed with a pallet truck rather than requiring a forklift. This may seem a minor point, but it allows a production operator to remove it from the warehouse rather than being dependent on availability of a forklift and driver.

When needed in the production area, a production person goes to the warehouse with a pallet truck and grabs the last pallet(s) in the lane. They sign the paperwork located with the pallet and drop it in the warehouse office as they exit.

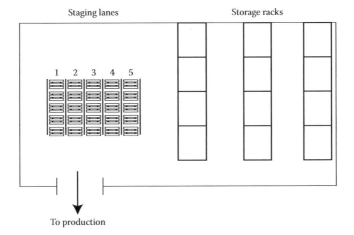

To production

Space is always a constraint in any plant and it may be hard to find space for prestaging. One of the benefits of increased flexibility from a lean changeover (LCO) program can be a reduction in inventory levels. This can free up warehouse space for prestaging.

Some plants have issues with their bill of materials (BOM). The BOM must be 100% accurate. Everything required to produce the product must be included in the BOM. In some plants, some components such as glue are treated as supplies. Rather than being in the BOM, they are managed using a 2-bin system or the like. The author once saw a plant shut down for 2 days because it ran out of the glue required to seal the shipping cases. Adding these materials to the BOM can help avoid this kind of disruption.

Documentation is another potential area for externalization following practices similar to those described above.

Many plants will need to perform a postproduction reconciliation and closeout of the previous order. This may be done on the production floor and may prevent starting the changeover until this is completed. It may be possible to delay this reconciliation either in whole or in part until the next run has commenced. Another alternative might be to move it off the production floor so that changeover can commence while the reconciliation is still being completed.

Changepart and Tool Handling

It should go without saying that all of these closeout tasks should be performed as much as possible while the line is still running. Forms should be fetched and any data such as lot numbers and machines used should be filled in ahead of time. In a process that produces bins of intermediate parts, it may be possible to count or weigh everything but the last bin ahead of time.

When changeparts, molds, dies, or other machine parts are required for the changeover, these should be prepared in advance. As they are returned to the storeroom after use, they must be inspected. Parts that were damaged in use should have been repaired before they were returned to storage, but this is not always done. Cleaning should have been done prior to returning to storage but may not have occurred. Before the parts are replaced on their storage racks, any required cleaning and repairs must be done so they are ready for immediate use.

In some industries, such as food and pharmaceutical, cleaning may have an expiration date. A part that was cleaned 15 days previously might need to be recleaned. If a calibration is required, this may have expired since last use. Damage may have occurred due to improper storage. For these reasons, it is important to pick all parts required for the next job while the current job is still running. If this is done, any discrepancies can be corrected before the parts are actually needed. All of this needs to be known, and more importantly, corrected in advance of the changeover. The time to discover any of these or other problems is most definitely not when mounting them on the machine.

Start with the tools required for changeover. Some of these may be cleaning materials, including brooms, mops, detergent, hoses, and more. These should be neatly and permanently organized on cleaning carts. The cart should be verified to assure that everything required is on it prior to end of production. It should be staged as close as possible to the production area. One option is to use wire rack carts of an appropriate size. They are open so the contents can be easily seen. Various shelving options allow configuration as needed. A valuable addition is a perforated pegboard that can be made into a shadow board. Most are familiar with these pegboards in a brown fiber material. This is probably not durable enough for most industrial usage. A better alternative is polyethylene pegboard available from most industrial supply houses. Be sure to get a wide assortment of hooks and hangers to allow maximum flexibility in design. Once a final design has been tested and found satisfactory, shadow outlines of the various brushes and so on can be drawn.

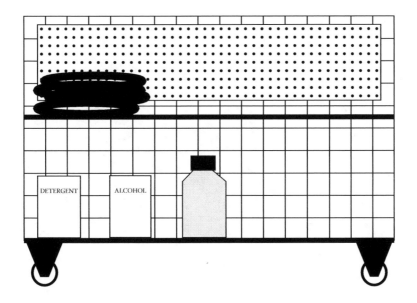

Tool carts for all tools required during changeover should also be provided. These carts should be limited to only the tools required. This facilitates organization and limits time spent rummaging through a tool box, perhaps located elsewhere in the room, for the proper tool. Mounting these on castors allows them to be easily moved where needed.

Proper racks for transport and storage are a must. They prevent damage that can occur from having parts jumbled on top of each other as is sometimes the practice. They provide additional benefits as well. A well-designed cart will allow near-instant visual determination that all parts are present. A well-designed cart will provide a visual aid to mounting the parts on the machine as well as help position them for easier mounting. Shown below is a cart for change parts for a rotary labeler. The front of the cart shows the starwheels and guides on a shadow board in their relation to each other as mounted on the machine. The rear portion of the cart has several slide-out shelves with assigned spaces for each of the numerous smaller parts. It also has a slide-out shelf for the rather heavy label magazine. This shelf is at the same elevation as the magazine mounting. This allows the operator to slide it from the cart onto the machine without having to lift it.

Photo courtesy of Septimatech Group Inc.

Carts will allow externalization of fetching the parts, but there may be some even better alternatives. This picture shows a centrifugal orienter running various sized bottles. A selector blade is the only changepart required to change between larger and smaller bottles. Rather than store this elsewhere, the machine builder provided a mounting in the orienter for the blade not in use. This is located about 6 inches away from the point that it is installed.

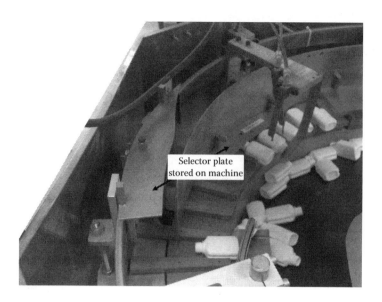

Several styles of cart can be useful in transporting and mounting heavy dies and molds. One type is called a T-cart or T-table from its layout. The die is loaded on the cross of the T in the storeroom, moved to the machine, and positioned so that the leg of the T is aligned to the die mounting. This can be done while the machine is running. Depending on the machine and the size of the die, the table can be on wheels with floor locks. Alternately, it may be moved on a motorized pallet jack or a forklift and docked to the machine. In some cases the T-table is permanently mounted at the machine and the die staged on it while the machine is running.

The table should have transfer balls or rollers to allow easy movement of the die. One alternative is to have transfer balls that lift with air. This allows compressed air to be connected to the table at the machine for ease of die handling. The compressed air is disconnected, allowing the die to sit directly on the table during transportation, preventing inadvertent movement of the die. Side guides are also recommended to prevent the die from inadvertently falling off.

If the die is especially large or heavy, it may be desirable to include powered rollers or a pneumatic pusher to aid in movement of the die between table and machine.

Once positioned and after the production run ends, the die is unfastened from the machine and moved up the leg of the T onto the empty section of the cross. The next die is now moved from its position on the other cross and down the T leg into the machine. After completing changeover, the table can be removed. If space or layout constraints preclude the use of a T layout, a rotary table can serve a similar function.

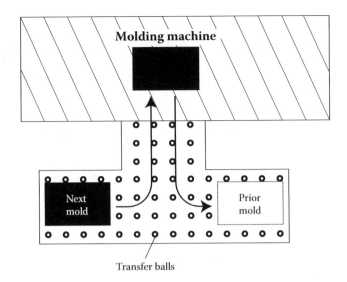

Transfer balls

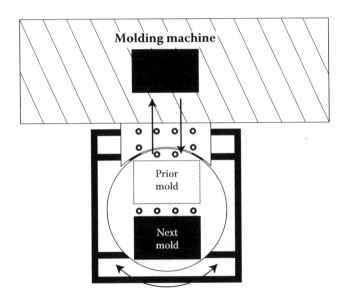

Another option is a bi-level cart. In the storeroom the die is placed on the lower shelf. The cart is moved to the machine and positioned in front to receive the die currently in use. At the end of production, the die is unfastened and moved onto the upper shelf. The two shelves are raised so that the lower shelf is at the elevation of the machine. The next die is slid into the machine, fastened, and production restarted.

The shelves are lowered (for safety) and the cart moved back to the storage room.

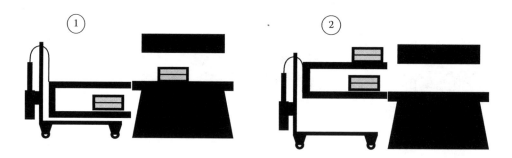

High-speed bottle-molding machines can use a dozen or more molds on a rotating turret. The individual molds may be brought to the machine on a pallet or in a cart. A mechanic then needs to unfasten and lift a mold out of the machine and carry it to the cart. They reverse the process to install the next mold. The process is ergonomically hazardous due to the weights and lifting involved.

A rotary cart and transfer bridge can speed up this operation. The cart has a rotating table with 9 storage positions (for a 8-mold machine). The transfer bridge allows the mold in position 1 to be unfastened from the machine and slid rather than carried to the 13th position on the storage table. The table is then rotated so that mold 1 is aligned with the transfer bridge. It is slid to the machine and mounted. The process is repeated until all 12 molds have been replaced. The cart is then rolled off to storage.

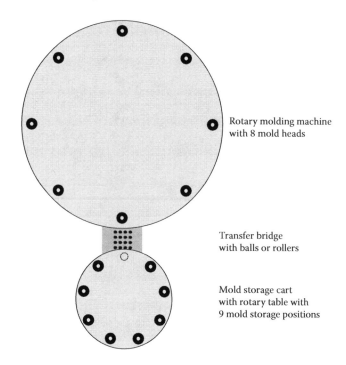

Rotary molding machine with 8 mold heads

Transfer bridge with balls or rollers

Mold storage cart with rotary table with 9 mold storage positions

Some may say that these examples are not really externalization because the actual die change can only occur with the machine stopped. This stems from the way one thinks about changeover. If the task is evaluated in its entirety (i.e., remove and replace the dies) they are correct. This obviously cannot be done with the machine in operation. If the task is broken down into its constituent components, there are many subtasks (e.g., bring the next dies from storage and stage at the machine) that can be externalized. This is why it is so important, when evaluating changeover, to break every task down into its smallest possible components. Each component can then be evaluated and improved. These are easy where it might be hard to identify opportunities when looking at the entire job.

Standardized Components

All parts that are product dependent must be standardized. This applies to molds, dies, timing screws, and change parts. Standardization does not mean similar, it must mean identical. All changeable parts must mount on the machine in such a manner that no adjustment is required. Standardization includes some obvious (though not always implemented) things like standard electrical and air connections. There should never be a need to rewire or replumb a machine between standardized parts.

Molds, dies, and similar components must also be standardized to allow mounting without adjustment. Shown below are two dies of different sizes.

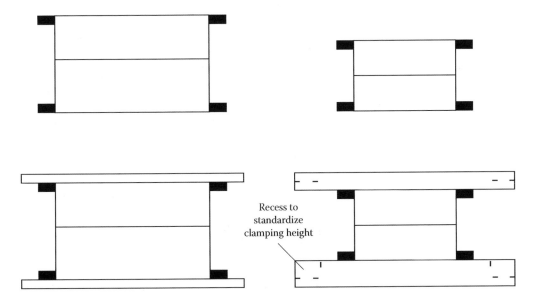

Recess to
standardize
clamping height

As built, the shut height of the machine must be adjusted for each. Different sizes mean that the opening height will be different for each, necessitating further adjustment. Different base sizes preclude the use of common locating guides. Different base thickness precludes the use of common mounting fasteners.

The need for these adjustments during changeover requires not only more time but higher skill levels. If all molds were the same size, they could slide into a location defined by fixed guides and quickly clamped in place.

One way to resolve this is by the use of additional mounting plates. As modified, the smaller die, mounted with two plates, is now identical to the larger die. The mounting plates are sized so that they can slide into common guides. The overall height is the same, which eliminates shut height adjustment, and the molds separate at a common height. Recesses are provided in the mounting plates so that the clamps are in the same position for each mold.

If the molds always run on the same machine, the plates may be permanently mounted. If they need to run on different machines, it will be necessary to mount plates or combinations of plates in preparation for each changeover. If they must be mounted each time, that must be done before the molds are needed at the machine.

Standardization allows an operator to quickly and accurately perform a mold change that previously required a skilled mechanic. The mechanic is now available to perform more valuable work elsewhere.

Pre-heating

In some processes, molds need to be heated before they can run and this can represent an externalization opportunity. This heating may take place after they are mounted using the normal heaters. In some cases, they are heated by being run with production. Heating this way will delay restart of good production and produce defective products that will need to be discarded or reworked at additional money and time cost. An additional risk can be caused by the time pressure of internal heating. Excessively rapid heating can damage the dies from uneven thermal expansion.

A better method is to heat the molds ahead of time. Depending on the mold and the temperature required, this might be done by circulating hot water or steam, by electric heaters either permanently installed or in blankets, or by gas flame. Shingo, in *A Revolution in Manufacturing*, mentions a plant that used the waste heat from an oven for heating dies.

If external heating is used, the hot dies will present a safety hazard during installation. Proper gloves and other safety and handling equipment must be provided to minimize risk to plant personnel.

Duplicate Components

Changes in the sequencing of the setup tasks can sometimes avoid lost production time as parts come up to temperature. The metal type and chase (type holder) in a hot stamp printer, commonly used to print a production/date code on a label or carton, can take 5–10 minutes to cool down after removal from the printer. While it would be possible to change the type while hot, it represents a potential safety hazard and the typical practice is to allow it to cool down first. Once cool, changing the type character is a fiddly task taking 5–10 minutes. In some plants, the set type must be inspected and approved before it can be mounted in the printer. After mounting, it can take another 5 minutes to heat to operating temperature.

The purchase of an additional type chase and additional type pieces can eliminate all of this. The next code can be set up and inspected (if required) before the end of the current production run. At the end of the run, the first step of changeover is to remove the current chase and insert the new one. The hot holder is set aside to cool down and the rest of the changeover on the labeling or other machine takes place while the new type comes up to temperature. After line restart, the chase is broken down, the individual type pieces are inspected for wear and replaced in storage.

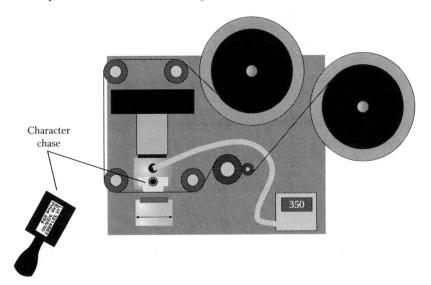

Character chase

A word on wear parts: The metal type is fairly durable but does wear and will need to be replaced periodically. Replacement frequency will vary depending on usage. Any problems due to the type must be noted during production and the type inspected after each use. This is not an area to skimp on, and worn type should be replaced immediately. It might be possible to postpone spending $1,000 on new type but at what cost? The worn type can be difficult to align properly, causing delays in setup. It may not print properly, causing further delays in production as well as rejected product. It is generally inadvisable to try to save money by only replacing the worn pieces. All should be replaced as a set to avoid slight differences that can cause print quality issues. This is true even when replacing type with identical new type from the same vendor.

Parts washing and ways to improve it are discussed in Chapter 5. There are certainly opportunities that must be pursued, but the key is externalizing cleaning. A filling machine will have wetted or fluid path components including hoses, pistons, valves, and nozzles that must be removed and cleaned between products. Improvements in the washroom will reduce the amount of time between removal and replacement but will still delay machine reassembly.

Additional parts sets should be purchased to allow washing to be externalized. The dirty parts are removed and set aside in a cart. The previously cleaned duplicate set can be mounted immediately. After production has restarted, the cart with the dirty parts can be moved to the washroom.

Externalizing washing in this manner has an important secondary effect as well. Reducing the time pressure in the washing process may also reduce errors and improve cleaning.

The picture below shows a liquid filling system specifically designed for externalization. The pump itself is driven via magnetic coupling through a stainless shell. This eliminates the need for any shaft seals, allowing the pump capsule to be completely sealed. The flow control valve (upper right above the nozzle) acts by pinching the tube closed. This eliminates the valve as a wetted part. For changeover, two thumbscrews release the pump and another opens the valve. The entire assembly, including hoses, pump, and nozzle, can be removed in seconds without disassembly. A previously cleaned and assembled pump, hose, and nozzle assembly is mounted in a few seconds more. The electronic controls allow all operating parameters to be selected from a menu, eliminating any setup adjustment.

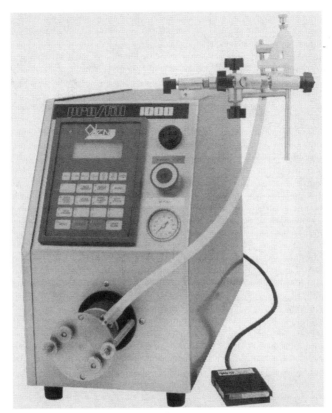

Photo courtesy of Oden Corporation

Some manufacturers take this even further and have entire duplicate machines or major components that can be quickly replaced. High-speed labelers take time to disassemble, clean, and reassemble after use. One alternative is to mount the labeling apparatus (head) on its own stand with castors. At the end of the production run it is wheeled off to a cleaning area. A second head is then wheeled up to the labeling machine where a mechanical docking station clamps it in the required location. A single umbilical jack supplies all power and control connections.

Photo courtesy of Krones Inc.

Some manufacturers purchase complete duplicate machines that dock on the packaging line. Tablet fillers are notoriously time consuming to clean and set up, and some companies have extra fillers with modular control systems. The mechanical portion of the filler is mounted on castors and has a docking station to allow it to dock to the conveyor in the same position every time. The controls, in a separate cabinet, are not duplicated and remain in place. This saves the cost of having a separate control panel for each filler. On completion of a job, the dirty filler is unplugged from the control panel, unlatched from the conveyor, and wheeled off to the washroom. A previously cleaned filler is wheeled into position, latched and plugged, and is ready to run.

It may not be necessary to have additional machines on a strict one-for-one basis. Depending on production schedules, a plant with 5 lines might only need 2–3 additional fillers instead of 5. In order for this to work, it is critical that all of them be truly interchangeable between lines.

Even small machines can be externalized with significant results. Some companies use a cold glue (think white library glue) to adhere a foil or paper tamper seal to bottles. The glue applicator consists of a reservoir and a roller than makes contact with the bottle neck, applying a thin film of glue. This is normally done on the main line conveyor. These gluers typically drip on the conveyor. As the reservoir is refilled, glue is sometimes spilled on the conveyor. This glue must be cleaned from the conveyor as it drips and must be cleaned daily from between the links. If not, it will not only contaminate the bottle bottoms but can eventually cause the conveyor chain to bind and break. Cleaning of the gluer online is time consuming and messy.

The author developed a conveyor system used in a number of plants to allow this to be externalized. A short belt conveyor is mounted along with the gluer on a wheeled base. A docking system is provided to latch the system to the main conveyor. Guide rails transfer the bottle from the main conveyor, under the gluer, and back to the main conveyor. Any drips fall on the short belt conveyor. The belt is a nonstick material that allows for easy cleaning of the drips. Motor, control, bearings, and other features are waterproof.

At the end of the production run, the conveyor and gluer is undocked and moved aside. A previously cleaned system is docked in its place. The dirty system is moved to a washroom where it can be washed with a pressure hose. It is then stored in readiness for the next line changeover requiring a gluer.

This externalization of the gluer had three benefits beyond reducing changeover time. It improved quality by reducing or eliminating a source of contamination of the bottle bottoms. Cleaning with a spray hose rather than buckets and sponges provided better cleaning. Finally, cleaning with a spray hose reduced the labor hours required for cleaning, freeing teammates for other tasks.

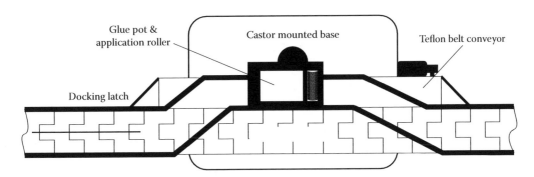

An even more substantial variation on multiple machines is multiple lines. One bottling plant had somewhat seasonal demand and had 7 lines in order to meet it. During most of the year they normally ran only 4–5 lines. When it came time to change products, they would move the operators to another, previously set up line. Then, with the other line stopped, the mechanics could do the changeover without interrupting production.

Sometimes it is not even necessary to have additional machines. Continuous inkjet coders have gotten much better over the years but are still messy due to overspray and underspray. Most plants with inkjets will have discoloration all around the inkjet area of use. Proper setup can reduce but will not eliminate the mess. One solution is to control it.

The typical installation puts the inkjet head next to the conveyor with no guarding. This allows any overspray to escape and collect wherever the aerosolized droplets may fall. A stainless steel box can be mounted around the ink head with a small hole in the bottom for the droplets to exit to the product. In another installation type, but a more problematic one, droplets are collected in the box. In this way very few droplets manage to find their way to the floor or the machine. At the end of production, the box is swapped out for a spare and taken away for cleaning.

Another plant that did a large-scale coating process had a problem with coatings contaminating large sections of the machine. They worked to reduce the extraneous amount of coating but were not able to completely eliminate it. They did the next best thing. They used a sticky plastic film to cover all flat and nonflat areas where the coating could accumulate. They run a high variety of coatings and do not need to clean every lot, so they let it build up a bit. At the end of the week or more frequently if required, the film and waste coating material is peeled off and discarded.

Duplicate machines are not a cheap solution and management may balk at spending the money. This is especially true if the justification is attempted on the basis of time saved rather than dollars saved. Management that is asked to spend $100,000 to save 30 minutes per day is likely to say no. If they are asked to spend that same money to save $5,000 per day, their answer will likely be very different.

Clean-in-Place

Some plants use clean-in-place (CIP) systems to simplify cleaning and make it faster and more repeatable. This is certainly beneficial, but if it takes place during changeover (internal) it still delays production restart. If CIP is to be used, it should be externalized if possible.

One design uses three permanently mounted pipes between the compounding room and the filling machine. Two pipes supply product to the filler and the third pipe is a return from the filler to the product room. When the line is operating, product is supplied to the filler via one of the two supply tubes. On completion of production, this pipe is disconnected from the filler and connected to the return pipe with a "U" fitting. It is possible to design a system like this using valves, but a removable U fitting with quick connectors eliminates the chances of the wrong valve combination being used.

After cleaning the filler, the pipe is connected at the filler and to the product reservoir in the product manufacturing room. On completion of changeover, the filler can be restarted and cleaning of the previously used pipe can proceed while production is ongoing.

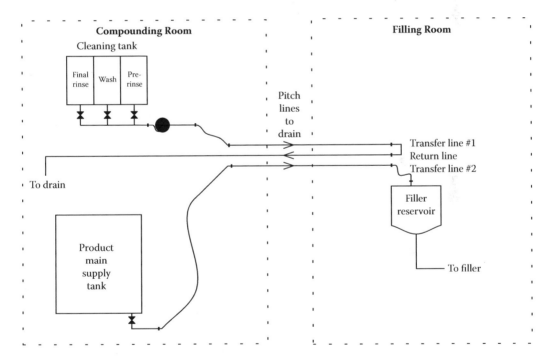

Dual Rollstands

Roll-fed machines should have dual roll stands so that the next roll can be loaded while production is ongoing. If an automatic splicing system is included, the actual changeover can be done "on the fly" with no machine stoppage. There are several designs for backup roll systems. The simplest extends the normal mounting stand and allows a backup roll on a shaft to be placed behind the in-use roll.

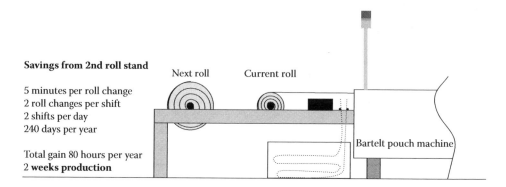

Savings from 2nd roll stand

5 minutes per roll change
2 roll changes per shift
2 shifts per day
240 days per year

Total gain 80 hours per year
2 **weeks production**

A variation that may be more useful for heavier rolls, such as metal coils or larger rolls of film or foil, places two mounting shafts on a rotating vertical turret. At the end of the previous coil, the turret rotates 180 degrees and the next coil is presented.

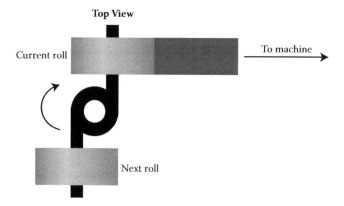

Top View

If very large rolls need to be handled, such as in a printing, papermaking, film extrusion, or similar processes, horizontal rotating design may be more appropriate. Two plates are mounted on a rotating axel. Each plate has two shaft supports. This allows the roll to be mounted on a shaft and then the shaft lifted into the two supports. These may be powered if needed to allow better unwind control. When it is time to change rolls, the assembly rotates 180 degrees, swapping the rolls front for back.

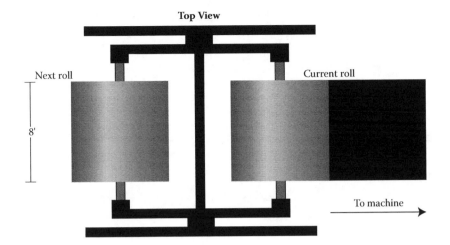

Backup film roll mounting has an advantage in production as well as in changeover. The author was once in a plant that lost 10 minutes, twice per shift (20 shifts per week), due to a lack of roll backup. An operator had to notice that the machine had run out and stopped, get a new roll from across the room, carry it to the machine, remove and disassemble the mounting shaft, mount the next roll, and splice it into the film tail.

Adding backup roll stands eliminated 5 minutes or more of downtime per roll. Over the course of a year this 20 minutes per day added up to 80 additional hours of machine production time.

Testing

Externalization does not apply only to cleanup and setup. It can also reduce startup time as well. Some processes will require certain operations to be completed after changeover but before starting production. Some even require that production be stopped after having restarted.

One plant coated adhesive onto 6-foot-wide rolls of plastic film. This process generates massive amounts of static electricity. As must be the case in all plants, safety was paramount for them. After each changeover, prior to starting production, an electrician was required to test the static discharge system with a megohmeter.

This normally caused a delay as the electrician was not called until changeover was finished. This led to an unacceptable situation of a machine waiting for an electrician. As noted in Chapter 2, the difference in cost per hour between even a skilled technician and production time may be a factor of 20 or more. Waiting should be eliminated or at least reduced by better scheduling. If waiting still must occur, it should be the electrician waiting for the machine, not vice versa.

In this case there was a further delay. Certain adhesives had a short shelf life so they did not mix them until they knew that the machine would pass the static test. Once the electrician had given the OK, there was a further 30–40 minute delay while the adhesive was prepared.

The question was raised about the cost of the adhesive, the cost of downtime, and the low probability of the machine failing the static test. It was decided that the daily loss of downtime outweighed the occasional loss of adhesive caused by static test failure. Based on this they decided to mix the adhesive ahead of time on the assumption that the static test would be OK. This allowed restart of production immediately upon completion of the test.

A similar situation occurred in an automated computer numerical control (CNC) machining plant making high-precision plastic parts. Their process was that after completion of machine setup they would run one piece. They would then stop the machine while the machinist used a coordinate measuring machine (CMM) to check all dimensions. During this inspection, 1–3 additional parts could have been made.

The machinists were very good at setups and the pieces very seldom failed the inspection. The cost of downtime outweighed the cost of the 2–3 rejected parts that might occasionally be produced. This was a good opportunity for letting the production continue while the first piece was inspected.

Conclusion

A manufacturing plant has one priority: produce quality product. It is only when it is producing that it is making (in the real sense of the word) money. Everyone must focus on maximizing available production time. Nothing that can be done with the line running should be done with it stopped.

Chapter 7

Execute

In an ideal world, after cleanup and setup are completed, it should be possible to restart the process and have it run immediately at full efficiency. This is the vertical startup that was discussed in Chapter 1. This startup time may be even greater than the time spent on cleanup and setup combined. In a few extreme cases, the production run may finish before the process is completely settled down, running at normal speed and efficiency.

There is a single cause of nonvertical startup: variability.

Some startup is caused by variability of the materials. Changeover improvement does not usually directly address material variability. This is usually the responsibility of the quality, materials, manufacturing, and other departments involved in developing and maintaining material specifications. One issue is that if changeover is not done precisely, it is difficult to determine where the issues lie. A finger-pointing game can develop with each side blaming the other. Manufacturing may say that precise setup is not possible because of material variability. The materials department may say that materials are in spec, and that the problem is that manufacturing is unable to set the machines properly. Each side is often at least partially right. Manufacturing must get the changeover under control. Once they do, and can show that they can perform them repeatably, time after time after time, they will be able to demonstrate the delays caused by material variability.

These delays need to be noted and quantified. They should be quantified in dollars if possible or in time if the cost per hour is not available. Once quantified, a tool exists to justify reducing material variability.

Most startup is caused by variability of the changeover. Setup is generally the main culprit but variability in cleanup sometimes causes startup delays as well. If the variability can be driven out of changeover, startup times will be dramatically reduced.

Execution is the key to driving out variability. Absent proper execution of the changeover, much of the value of the elimination, simplification, and externalization will be lost. The improved changeover must be executed exactly.

Proper changeover execution depends upon several factors. Well-written and comprehensive standard operating procedures (SOPs) and checklists are one. These were discussed in detail in Chapter 3. Proper supervision of changeovers to assure that they are properly done is another. Quantifiable, repeatable, exact adjustment is one of the most important factors and is the focus of this chapter. Ideally, it should be possible to set everything to the target set point and run. In reality, although much variability may be driven out of the setup, some will still remain. SOPs must be written to give latitude to adjustment so they can be tweaked as needed. When tweaking is required, the actual run setpoint must be recorded in an equipment log or database. Where the reason for the variance from the SOP is known, it must be noted. If the variance becomes consistent, perhaps because of machine wear, this should generate either a maintenance action to bring it back to ideal or a revision of the SOP to reflect the new required settings

It is critical to measure and standardize the elements of changeover. Equally critical is the measurement of the changeover itself. Every changeover must be measured and recorded. This should be done graphically with trend lines shown. Discrepancy reports need to be generated when changeover takes significantly more time than normal. The discrepancy report must be investigated and the reasons for delay corrected as necessary. A discrepancy report should be also generated when the changeover takes less time than normal. If a normal 4-hour changeover is finished in 3, it may be that something was not done that should have been. This must be corrected. In other cases it may be that the hour was saved because there were no delays and everything just went right. This should be noted as well since it may give clues to permanently improve the changeover.

One plant had different departments running almost the same processes on the same equipment. Department 1 took 30–35 hours to do a major changeover, mostly consisting of cleaning. Department 2 did the same cleaning but normally took 20–25 hours to complete it. When this was first discovered (after it had been going on for years), the first thought was that Department 2 was not following the SOP or took shortcuts. Closer evaluation showed that they were following the same process, they were just better organized and more effective. A significant reduction in changeover time was attained just by transferring Department 2's methods to Department 1.

Many plants will have different types or levels of changeover. A food canning plant may can beans for several different brands. There may be changeovers where the only change is the label. A pharmaceutical plant may package the same tablet in several different size bottles. Changeover in this case may involve changing the line but with little or no cleaning. Then there may be the major changeover.

A paint plant may be able to change from white to red with little more than flushing of the fluid path. Changing from red to white may involve substantially more work to avoid giving a tint to the white paint.

It is not realistic to compare changeover times when the changeovers are very different in nature. One solution is to develop categories of changeover such as minor, major, color, brand, size, and perhaps classifications between when required. The various categories should be tracked and compared against one another. Tracking different types of changeovers may also give clues to reducing changeover times by converting a major changeover to a minor changeover.

Food plants that produce products with peanuts and other nuts must be extremely careful to avoid cross contamination of products without nuts. Cleaning when going from nut to non-nut products must be considerably more rigorous than the reverse. One way to reduce changeover time is to schedule the non-nut products during a specific period, then run the nut products with minimal cleanup time between them. The schedule might be Monday through Tuesday non-nut production and Thursday through Friday for nut production. Wednesday is used if necessary for non-nut and for the minor interproduct cleaning. The weekend is used to perform a major cleaning so that the non-nut products can be run again on Monday.

Opportunities for these kinds of improvements may remain hidden absent proper classification and tracking of all changeover times.

Measurement of changeover times is not always easy. Clocking the time between when the line stops and restarts can be fairly straightforward, but is only a portion of total changeover time. The slowdown period, where it exists toward the end of the run, must be measured. More importantly, the startup period must be measured. By definition, startup ends when the line is running at normal speed and efficiency. Some lines will have an automated data collection system that shows a throughput graph. This allows the point at which throughput stabilized to be seen fairly easily. If this type of system is not present, it can be harder to measure. One technique is to set a period of time, say, 10 minutes (though it could be 5 or 15), during which the line runs continuously with no slowdowns or stoppages. Once the line has run continuously for this period, startup is considered complete.

One plant developed a different approach. They determined that normal production speed was approximately 4 cases per minute. They considered startup time as ending when they had completed 100 cases or about 25 minutes after restart in ideal conditions. Measuring the length of time between restart and completion of 100 cases did not give them the absolute startup time. What it did give them was an easily collected and understood metric that allowed them to compare relative startup times across changeovers.

It may be possible to perform changeovers and setups with no adjustments necessary. Some machines are designed with changeparts, jigs, or fixtures where all adjustment is built in and they have only to be mounted in a fixed position. Most machines will have some combination of adjustment and changeable parts. Where adjustments are required, they must be done exactly. This requires specification in the SOP of all the proper setpoints. It also requires physical means to set part positions.

It is not acceptable to say "Adjust the guide so that it is as close as possible to the part." This requires judgment on the part of the setup person. One might understand *close* as meaning 1/16 inch. Another might understand that 1/4 inch is appropriate. While it is possible that both are wrong, it is not possible that both are right. The SOP must include the specific distance, for example, "Adjust the guide so that it is 1/4 inch from the part." The question then is how to measure that distance. Proper tools and devices must be provided. This chapter will focus on techniques for achieving repeatable adjustments.

As a general rule, the measuring device should be permanently mounted to the machine in the position it is to be used wherever possible. Earlier, this book discussed the problems with tools and why it is desirable to eliminate them wherever possible. This applies to measuring devices as well. A gauge that is inserted between two parts to set spacing is a tool with all the issues that any tool brings. It is far better to have a permanent stop mounted, set to the proper distance.

Digital measuring devices are always to be preferred over analog. This applies to physical measurements as well as speed, temperature, pressure, time, and other parameters. Analog devices require the setup person to make a judgment. "Does that pointer indicate 9 1/2 or 9 5/8?" "Does that temperature control indicate 150 degrees or 155?" The digital indicator will provide a single number that is not subject to interpretation or misreading.

Speed

Speed is one of the most critical parameters in most processes, yet it is also the one most likely to be ignored. When not ignored, it may not be measured accurately. Many people do not seem to realize that proper machine speeds, even of something seemingly as simple as a conveyor, are critical to proper manufacturing performance. Speed is critical with even a single machine. When multiple machines, working together, are not running at the proper speeds the effect is multiplied. Improper speed settings will cause the line to be imbalanced. An improperly balanced line will not run smoothly, causing excessive stress on equipment, more need for operator intervention, possible quality problems (e.g., insufficient dwell time in a process), and will usually have a negative impact on production capacity.

Speeds may be set by eye, feel, and occasionally by ear. Some speeds are controlled by a mechanical slide base and spring-loaded pulleys. Others are controlled electronically by a single turn knob graduated from 0–10 or 0–100. The scale and pointer may be difficult to align precisely. Even when they are, speed may not be linear. A very small adjustment of the knob may result in a relatively large speed change. In one case, an SOP called for a conveyor speed to be set at 6.5 on a 0–10 scale. This required setting the knob about halfway between 6 and 7 and a mark had been made to indicate it. In this case, the actual practice was even worse. The mechanic would set the conveyor speed by eye, with each mechanic having a different opinion of what the best speed was. Once set, they would loosen the set-screw holding the control knob and reposition it to indicate about 6.5.

A tachometer mounted on the conveyor, displaying chain speed (feet/minute) solved this problem. The tachometer allows direct measurement of the critical parameter. Proper measurement allows proper setting. Its digital display allows it to be visible to everyone.

Some plants may use a handheld tachometer for setup. This is much better than nothing as it helps ensure that the speed was at least correct at one point in time. A handheld tachometer has several drawbacks though. First, it is a tool. It can take time to locate it and it may not be available when needed. Second, there is some technique required for its proper use. If not used properly, it is possible for two people to measure two different speeds. Finally, it only gives the speed at the time of measurement. It does not provide speed during operation when process loads or operator tinkering may change it.

Fixed digital electronic tachometers are readily available at very economical pricing, usually less than $300 or so. A toothed gear is mounted on the motor or other rotating component and a magnetic pickup senses the teeth as they rotate past. Internal batteries with 2- to 3-year lives can make installation even simpler by eliminating the need for electrical installation. In short, there is no excuse not to mount a tachometer on any piece of equipment that moves. This is true for fixed-speed machines, but is especially true for variable-speed machines. Electronic tachometers usually have some rudimentary programming capabilities. This allows them to be programmed to read the most useful metric. A tachometer on a conveyor should display linear speed, typically meters or feet per minute, rather than motor or pulley revolutions per minute (RPM). A stamping machine or a bottle capper should display parts per minute and so on. For other machines, RPM might be the most appropriate metric.

Speed can change as the load on the machine changes. A machine set to 100 CPM (cycles per minute) during setup under no load may slow to 95 CPM under manufacturing conditions. Several manufacturers offer controllers that incorporate tachometers. The desired speed is set in the controller and it varies the power to the motor as required to keep the speed constant.

Positioning

Proper positioning of adjustable machine components is critical for proper operation. There are a number of techniques and tools to assure proper position adjustments.

Scribe Marks

One of the simpler and more common techniques is scribe marks. The mechanic marks the various positions where they have found the best operation. This may be done by scribing or scratching a line on the machine, or they may be marked with a marking pen. In general, this practice is not recommended for several reasons. First, the mark is made where one person thinks best and may not actually represent the best setting. Another objection to the practice is that it may not be clear which mark pertains to which product. As new products are added to the mix and old ones discontinued, setpoints, some useful and some obsolete, accumulate. Setpoints will change due to machine wear or maintenance and many of the marks will become obsolete. Finally, these handmade marks on a machine just look messy. Aesthetic appearance is important to proper functioning of a machine. Cleanliness may have no direct functional impact, but operators and mechanics will simply take better care of a machine that has a clean, well-maintained appearance. As they take better care, the machine is cleaner and better maintained. Virtuous cycles are the very best kind.

In one plant they overcame the issues of scribe marks by printing them on paper labels that were then mounted to the machine. This seemed to work well, but they did require periodic replacement as they became worn, dirty, or damaged.

A variation on this theme is engraved metal plates. These can work best when there is a stable mix of relatively few products. A plate is engraved with each position clearly marked and fixed to the machine. A pointer on the other adjusting component indicates when the proper adjustment is achieved. One option is to use a fork instead of a pointer. This machine is adjusted until the appropriate indicating mark is centered in the fork. This allows a larger, more visible mark to be used.

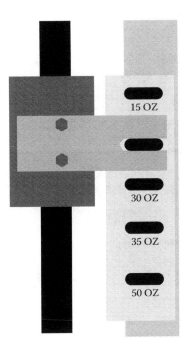

Scales

Teammates sometimes use scales or tape measures to determine correct position during setup. These are tools with the associated issues as discussed in Chapter 5 and should be avoided if possible. It may not always be clear exactly where the measurements are to be taken, causing additional variation. Scales and rulers are handy, but for setup they must be mounted permanently on the machine. They may be mounted on either the fixed or moving component, with the pointer mounted on the opposite component.

A variety of scale types are available. One style is printed on a plastic film with a self-adhesive backing. These are available in a variety of sizes, styles, and gradations to match specific applications. They are also available in arcs and dials for rotary or angular measurement.

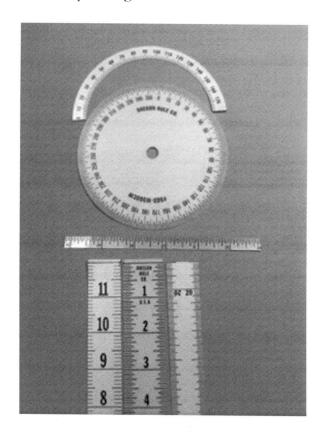

Stainless steel scales are available from tool supply houses in a variety of configurations. They can be mounted with epoxy or rivets. One drawback is that many have both metric and inch gradations. If used for machine setting, this is likely to cause confusion. When purchasing scales for this kind of application, be sure to purchase scales that read only inches or metric but not both.

Ambiguous scale-combines fractional inch and metric

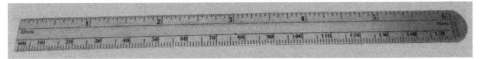

Another issue to consider with scales is the unit of measurement. Most inch scales are divided in 16ths, 32nds, or even 64ths. The plethora of marks can make them difficult to read even in the best of conditions. Once mounted on the machine, they are seldom in the best of conditions. Dirt and scratching may obscure the marks. Poor lighting may further exacerbate the problem. These difficulties and the need for interpretation are likely to lead to error.

It is preferable to use decimal scales. These are available, though sometimes hard to find, for foot/inch scales. They are readily available in metric, which is naturally decimal. In this type of application, the scales are being used for position indication rather than for measurement, so the choice of inch or metric is unimportant. What is important is that the setup person can easily read the target setting. Avoid using scales that have finer measurements than required. If setting to a tenth of an inch is required, using scales marked in hundredths can add difficulty and confusion.

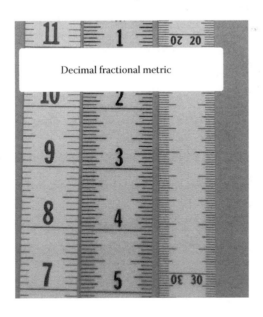

Decimal fractional metric

Whichever is selected—inch or metric—standardize throughout the line. Using metric scales on some machines, fractional inch scales on others, and decimal inch scales on still others is a recipe for confusion.

When using plastic scales, especially in exposed locations, it is a good idea to mill a slight recess where the scale is to be mounted. This will help protect the scale from scuffing and other damage.

A scale, by itself, is not much more than a decoration. In order to be effective, it must be combined with an effective pointer. Pointers are not always all that they should be. Two of the most common issues with pointers are positional instability and parallax.

Pointers must be made of rigid material. Anything mounted on a machine is subject to being knocked about and pointers are no exception. They must be tough enough to withstand this normal wear and tear without bending or flexing. Once they are bent or moved, their positional value is lost until they are reset and recalibrated. Getting slightly bent is even worse. They may continue to be used without anyone recognizing that they are no longer correct. Pointers should have no adjustability once mounted. In at least one case, a machine builder mounted a pointer with slots to allow calibration adjustment. Their heart may have been in the right place, but it is unlikely that the pointer would long remain in the right place. One must always assume that anything that is adjustable eventually will be adjusted, often unnecessarily and incorrectly. The scale/pointer combination does not need to be calibrated. Once the scale and pointer have been mounted, optimal machine settings are determined, usually by trial and error, and the readings for those settings recorded. It doesn't matter whether that reads 1.5 or 2.6 inches. What does matter is that whatever the reading for that setting, returning to that reading will return the machine to that position. Rigidity of pointers and scales cannot be overemphasized.

A second issue with pointers is parallax error. Parallax error occurs when the pointer is read from an angle rather than directly in line with the scale. If the line of sight between eyeball and scale is not perpendicular, the apparent reading from the pointer will be incorrect. Part of this issue is training. Operators and mechanics must be trained to read scales properly. Pointer design can also help. Some pointers will have their face painted black and the sides red (or other color combination). When reading the scale, only the black face should be visible. If red is seen, it indicates that the pointer is being incorrectly read. Some systems add a mirror to the scale to achieve a similar effect. When reading the pointer, it must be aligned with its mirror image. If the two are not aligned, it is not being properly read.

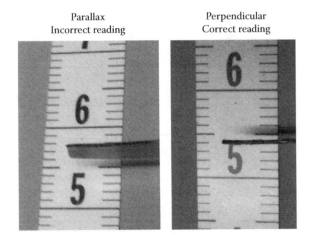

Parallax
Incorrect reading

Perpendicular
Correct reading

Parallax is aggravated by any distance between pointer and scale. It is possible to correctly read a scale where the pointer is a half inch away, but the distance makes it much more difficult. Training will help but is far from foolproof. Put the scale and pointer as close together as possible and even if there is some reading error, it will be minimized.

Plastic magnifying pointers are also available for fine work. These are circular or rectangular lenses with a fine line scribed in their center. They are mounted so that they ride on or just above the surface of the scale. They magnify the dimensional marks, making them easier to read. Parallax error is virtually eliminated by the proximity of the pointer and scale.

Some machine builders use scales that are meant to be aligned with an edge of the movable component. This can work well if there is a close alignment of a component feature and scale. If this technique is used, the edge to be read against the scale must be clearly marked. As with any technique, it is possible to use it incorrectly. The following figure shows an excellent though flawed application of the concept. A bar running the length of the machine has a scale mounted to it. Various brackets are set against the scale with the bracket edge replacing a pointer. In this case, there is no indication to show whether the proper reading is on the left or right side of the bracket. This is easily corrected by adding an arrow to the appropriate side of the bracket. An arrow could be drawn with a marking pen but might fade with time. A cold chisel can be used to make a simple "V" at the side of the bracket for a permanent mark.

Gauges

Gauges are tools and should be avoided. The exception is when the gauge is permanently mounted. When permanently mounted, they provide fixed stop positions. There are a number of ways gauges can be incorporated into changeover.

Thickness Gauge

Perhaps the simplest use is for something like an induction sealer, capper, or other machine where the distance between product and machine is constant regardless of product size. If an induction sealer head is to be 1/4 inch above the top of the product, a 1/4-inch-thick piece of plastic may be used to verify this spacing. A product is placed under the head, the gauge laid on top of it, and the head is lowered until it touches the gauge.

An induction sealer can be 18 inches long or more, which can make it difficult to get parallel to the conveyor. The distance between the head and the top of the bottle is always the same, which lends itself to the use of a gauge. In this case, a piece of plastic of the appropriate thickness is cut to the same length as the sealer head. This allows two bottles to be placed under the head, one at each end, and the gauge is laid across them. The head is lowered until it touches the gauge, then raised just enough to slide the gauge out. If the head does not touch the gauge at both ends, it indicates that the head is not parallel and needs to be adjusted.

When a gauge like this is used, it must be attached to the machine via a keeper wire. A hook, bracket, or other means of consistent storage at the point of use must also be provided.

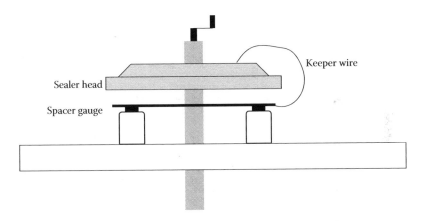

A variation on this can be used for setting conveyor rails and similar equipment. Plastic pieces are cut to the appropriate length for each setting. Rather than move these from bracket to bracket, multiple copies of each gauge should be made and hung at each bracket. As always, these should be clearly marked and color coded by product. Depending on the bracket style, ½-inch (or other suitable diameter) PVC plastic pipe can be a useful material for these gauges. The pipe is cut in half lengthwise and then cut to length forming half-round gauges. A similar system is commercially available with specially extruded plastic that, when cut to length, snaps over the bracket.

Whenever gauges are used, any clearances, either positive or in the case of grippers, negative, must be built into the gauge. The teammate performing the setup must be able to bring the machine component into contact with the gauge and not have to worry about whether it is too tight or too loose.

Some machines have adjustable parts that are subject to normal wear. An example is the tightening wheels on the friction wheel capper. The scales used to set the spacing between them may be fine when the wheels are brand new, but they are a soft rubber and wear over time. As they do, the scale setpoints no longer serve. A gauge may be the most appropriate tool in this case. The following figure shows a gauge made from a plastic block about 6 inches long so that it can be set on the conveyor. Slots in the side are cut to slightly less than the bottle diameter. When the side gripper belts are adjusted in against the block, they have just the right amount of pressure to correctly grip the bottle. The top of the gauge is sized slightly smaller than the cap. When the wheels touch the gauge, they will have just the right amount of pressure on the cap regardless of reduction in the wheel diameter due to wear.

Belt and wheel heights are set to the slots and shoulders on the block.

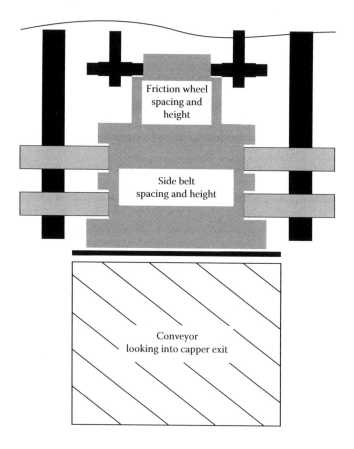

Leaf Gauge

If there are only two products to be run, or at least only two set positions, mechanical stops are easily mounted. The teammate only need loosen the component, move it all the way in or out against the stop, and retighten. In most cases, there will be more than two positions, making this technique more complicated though still feasible. There are several alternatives that can provide the same effect for multiple positions.

A leaf or jackknife gauge consists of a number of flat metal pieces, each of the appropriate length. They have a hole on one end that allows them to be permanently mounted on a bracket with a pin. In use, the appropriate leaf is swung into position and the machine components brought into contact.

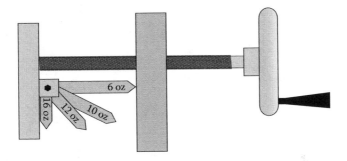

Another alternative is a rotating stop block. This consists of a circular block mounted to the machine with a central bolt allowing rotation. On the periphery a number of rods are mounted. Depending on the application, the rods may be either perpendicular or parallel to the axis of rotation. These rods are cut to the appropriate length for each setpoint. In some cases, bolts with locking nuts may be used instead. In both instances, the mounting block should be thick enough that some adjustment is possible, both for initial calibration and to compensate for machine wear. In both cases, the rods or bolts must be firmly fastened to discourage adjustment when not absolutely warranted. One way to do this is to use a double setscrew. Once the rod or bolt has been properly positioned, a setscrew locks it in place. A second setscrew is then placed on top of the first. This not only prevents the setscrew from vibrating loose, the unauthorized adjuster may not realize that after loosening the outer screw there is a second one underneath. Hopefully they will then give up.

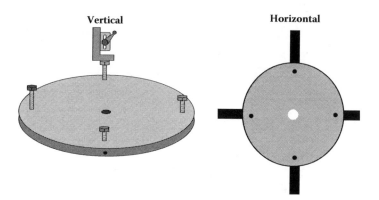

A variation on the rotating block is a sliding block. This is a square or rectangular block mounted on pins so that it can slide linearly. Rods or bolts of the appropriate lengths extend out of one side. The block is mounted across the path of travel of the moving part. It is slid into position and the moving part brought up to touch the appropriate stop.

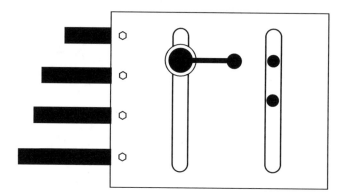

Digital Indicators

Digital indicators are mechanical or electrical devices that provide digital position readouts. They may be used to indicate rotational, angular, or linear position. They are often mounted on cranks or knobs on an adjusting screw. As the screw is turned, the indicator counts up or down until the proper setpoint is reached. As they are digital, they are not subject to interpretation. Since they can divide each revolution into hundredths, they are very precise. They are available in both mechanical and electronic versions.

Mechanical indicators mount on a rotating shaft. The housing is fixed to the machine. As the shaft is rotated, it drives a series of gears in the indicator that culminate in a series of numbered dials.

Photo courtesy of Siko Corporation

Under normal use, the internal mechanism is rugged and should last indefinitely. Care does need to be exercised in selection and mounting, though. This is a delicate precision tool and must be treated as such. It should not be exposed to excessive shock or vibration. Most importantly, it must ride on the adjusting shaft and must never be used to transmit force between crank and screw.

Analog versions are also available for mounting in the face of adjusting knobs. These have a dial and a pointer. An internal weight remains in a fixed position as the knob is turned operating the counter.

Photo courtesy of Siko Corporation

When linear motion must be measured, there are several options. One option uses a digital mechanism similar to that discussed previously. It is mounted in a housing with a spring-loaded retractable wire. As the machine is adjusted, the wire is pulled out, rotating the indicator showing the total linear travel.

Electronic digital indicators have several advantages over mechanical versions, though they have the complication of requiring a power source either via battery or wired connection. One major advantage is that many of them can provide remote as well as local readouts. This remote feature makes this type of indicator particularly useful when it might otherwise be difficult to read a local readout.

The output can be used as input to a programmable logic controller (PLC) or control system to block operation if the setting is not correct. Some models can receive a signal from the PLC to indicate what the setting should be. Rather than indicating the setpoint value, it indicates the positive or negative deviation from the target value. The teammate turns the adjusting screw until the indicator reads zero. Red and green light-emitting diodes (LEDs) may be added for additional visibility. A quick glance at a machine can show whether all settings show green and are correct. An incorrect setting with its red LED will be easily noticed.

Setpoint indicator light

Photo courtesy of Siko Corporation

Anyone who was worked with an electronic caliper will know how precise and reliable they are. Electronic digital indicators function similarly. A magnetic strip is typically mounted on the fixed machine component. The electronic pickup is mounted on the corresponding component. As it moves over the tape, it reads linear distance traveled. Highly precise, 0.001 inch or better, settings are routine.

Photo courtesy of Siko Corporation

Micrometer Adjusters

Some adjustments require very high levels of precision. Micrometer screws, either mechanical or electronic, can provide precisions to better than 0.001 inch. An advantage of the micrometer screw is that it not only provides extremely precise positional indication, the screw itself also makes the very fine adjustment. In the example shown, there was a need to adjust a selector guide to discriminate between components that were only a few thousandths of an inch different in diameter.

The system shown uses a vernier rather than a digital indicator. One disadvantage of this vernier type of micrometer is that it can be hard to read and requires skill to interpret correctly. The advantage is that it is extremely rugged. This was a major consideration in this case as it was mounted on a vibratory bowl feeder subject to continuous and extreme vibration. In another, less extreme mounting, it might be preferable to use a digital micrometer screw.

Adjusting Stops and Jacking Screws

Precise repeatable measurement by itself is not enough. It must be easy to adjust the component to the required position. Best of all is a stop, or a series of stops such as the rotating gauge. The component is pushed against the stop and locked in place. This is not always possible. In some cases moving the component in small, controllable, increments is difficult to do. A sliding base with a screw can be the solution. Commercially available linear actuators combine screws, slides, and mountings with other components to allow virtually any configuration for 1-, 2-, or 3-dimensional movement.

It may not always be feasible to add a proper slide base and adjusting screw to a machine. A simple alternative that can be fabricated in-plant is a jacking screw.

The following figure shows a machine with a slotted sub-base bolted to a main base fixed to the floor. During setup it is necessary to make small adjustments to the machine position by loosening the 4 bolts in the slotted holes, aligning the machine, then retightening the bolts. The problem was that the movements required were very slight, on the order of 1/16 to 1/8 inch, and this was hard to achieve.

The problem was solved with 4 short segments of angle steel. These were drilled with holes for mounting to the baseplate and a threaded hole in the vertical portion for a jacking bolt. Once mounted it was easy to move the machine precisely in small increments by loosening the hold-down bolts, loosening and tightening the jacking bolts to achieve position, then retightening the hold-down bolts. In some circumstances 2 jacking bolts might have been enough. In this case, there was a need for very precise angular setting so 4 bolts, 1 on each corner, were used.

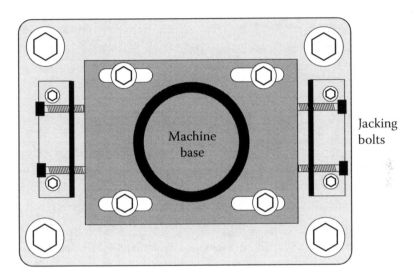

Jacking bolts

Use the Force

Virtually all machine adjustments will have some play in them. In a well-designed and maintained machine this play will be minimal but will still exist. This means that the result of adjusting a component up to a setpoint will give a slightly different result than adjusting the same component down to the same setpoint. Sometimes it is not even noticeable at setup. A capper head may be adjusted precisely at setup. As it was adjusted down from a larger previous product, there can be some residual play. As the machine runs, natural vibration can cause it to move down slightly.

If all adjustments are made by raising the head to the proper height, there will be no downward slack. The elevating screw will have taken it up. Upward slack is of little importance because there is nothing trying to push the head up. When going from a shorter to a taller bottle, this is not an issue. When going from a taller to a shorter bottle, the head should be lowered past the setpoint and then raised up to it.

In general, all adjustments must be made against the natural forces involved to eliminate the imprecision caused by slack in machine components.

The goal of lean changeover is to reduce changeover elapsed time, including startup time. A fast but imprecise changeover will be at the expense of longer startup time and more rejected or reworked product. Perhaps even worse than rejected products, are marginal products that are shipped. The customer doesn't know that they are within specification. They only know that they are not quite like the last time. It is better to perform a precise changeover and produce good products. The extra time spent on getting the setups correct will be paid back with reduced startup time and overall reduced total changeover time.

Conclusion

There are two keys to precise setups: good SOPs and checklists must be rigorously used. Failure to use them every single changeover will result in nonstandard changeovers. Good SOPs are not enough by themselves, either. Proper low and high technologies must be deployed so that every teammate can perform setup precisely the same way time after time after time.

Chapter 8

Develop and Implement a Program

It is not enough to know how to improve and reduce changeover. A program must be developed and implemented. How the program is implemented is perhaps the major factor that will determine success or failure. Any lean changeover (LCO) program must be carefully thought out in order to achieve maximum buy-in by the participants. Absent buy-in, no program will be successful.

The first person, or group, who must buy in is top management. Without management support, the program cannot succeed. With support, it is hard to fail. Everyone in the plant does, or should do, what is expected of them by their boss. The boss's priorities become the subordinate's priorities, and this is how it must be for any organization to run smoothly. If LCO is a priority for management, it will be a priority for everyone. This support will take several forms. One is managerial or moral support. They must let everyone in the organization know that the LCO program is important to the company and thus important to all the people in the company, even if they do not see a direct connection.

As a first step in any LCO program, management must have a clear idea of why they are embarking on the program. There are a number of reasons, including increased throughput, faster response to customers, more flexibility, better capacity utilization, reduced costs, or a combination of all of these. In some cases, the reason may be as simple as an imposition from the corporate hierarchy. The reason for the program is important because it will influence the focus of the LCO efforts. If the reason is not clearly understood, it will be difficult or impossible to set goals, and an unset goal is an unachievable goal.

Management must be ready to direct other departments to provide support to the LCO team. The finance department must be directed to develop cost models (discussed in Chapter 2) to allow financial justification. Purchasing may need to be directed to purchase higher-quality/less-variable materials. Engineering may need to be directed to modify equipment or facilities. Other departments will need to be directed to provide support as well.

Members of the LCO team will need to be relieved of some of their regular duties to allow them time to participate in the LCO program. This will include periodic, perhaps weekly, team meetings, training courses, and workshops, as well as time to develop and implement the improvements. Absent management support, immediate supervisors may be reluctant to allow this. Their goal is to get production out the door. If their people are taken away to work on changeover, this can interfere with this goal, at least in the short term.

Finally, management must provide material support to the changeover team. This can include funding for books and instructional materials, on- and off-site training, benchmarking visits to other plants, attendance at trade shows, and the like. It may be hard to show a direct return on this type of expenditure. Management must provide a reasonable level of funding for it without requiring a direct cost justification.

As improvements are identified, the team will need money for implementation. Much of this will be relatively minor expenses that may be available from operating budgets. Specific cost justifications may or may not be necessary, depending on the amounts and company policies. Other improvements may require larger, capital, expenditures. The team must be expected to provide financial justification for all of these expenditures. Management must be disposed to support these expenditures when they are justified.

Process or Project?

There are two ways that any improvement program can be implemented. Both have the same goal but different approaches. One is a process or open-ended approach which often goes under the umbrella of "continuous improvement." The basic concept is that the program will go on in perpetuity. The drawback to this approach is that nothing ever continues in perpetuity. Team members come and go. They remain on the team but lose interest. Other competing goals are introduced. For these or other reasons, the program sooner or later runs out of steam and dies.

Many who have been in industry have seen this happen with a number of programs that are implemented, theoretically in perpetuity, and then gradually fade away. In K–12 education, teachers even have a derisive acronym for it: TYNT (This Year's New Thing). People who go through several of these cycles usually become cynical. They will not invest themselves too deeply into it because, despite assurances that "This time we really mean it!," they suspect that it will not last.

An open-ended program often has no concrete goals. It may be expressed as "reduce changeover times," but by how much and by when? This can undercut any sense of achievement by the team members.

An alternate approach is to treat LCO as a series of projects. As a project, the program is begun with a specific objective in mind, such as reduce changeover time by 50%, and a specific deadline, such as 6 months. It can focus on a specific machine or line, or it may focus on a specific element of changeover, such as getting materials from the warehouse to the line in a timely manner. It might focus only on developing a complete set of standard operating procedures (SOPs) and checklists, or developing a system for measuring changeover time.

In short, it must have a specific focus, goal, and timeline.

At the end of the time allotted for the project, it will be evaluated. Was everything accomplished? Were the goals met? What was the result in terms of changeover time? What were the costs and benefits (in dollars)? What went right, what went wrong, and what could have been improved?

Depending on what was accomplished versus what was undertaken, the team should be recognized. This may be something as simple as a memo from management. It might also take the form of a recognition poster on plant bulletin boards or coverage in a company newsletter. Management might treat the team to a pizza session to discuss the project, results, and next steps. There are many ways to recognize achievement. The most appropriate way will depend on the people involved, the company, and the project.

The key is that there is recognition. It is fine to say that people should be self-motivated. They should be and many are. Even the most self-motivated person still gets a boost from external recognition. It may "go without saying," but say it anyway.

While the LCO project is ongoing, determine what the next project will be. It could simply be a continuation of the existing project or it may focus on different issues. It may be the same team members. If the focus of the new project is different or if some members have lost enthusiasm or moved to different positions, it may be a different team. It should be prepared in advance so that the day after one project is completed, another can be immediately begun. Do not allow any delay between projects. Delay is likely to break the momentum and make it harder to get started again.

The process-versus-project approach is not an either/or but a continuum. Both approaches can be combined as is most appropriate for a given set of circumstances.

Regardless of the approach chosen, the LCO program must be ongoing. There are so many opportunities that no plant will ever get to the point where they can say that they are finished. There will always be further opportunities.

Forming the Team

Every department in the plant has an impact on, changeover and every department should be represented on the LCO team. Everyone must realize what their impact on the project is and be prepared to provide support as necessary. Each department should be required to designate a specific member to the LCO team. Depending on the focus of the project, some of these will be deeply involved. Others may only need to provide occasional advice and support. Making them members of the team identifies them as the contact or go-to person for questions and resources pertaining to their area.

The Merriam-Webster dictionary defines the term *champion* as "militant advocate or defender of a cause." It is an appropriate title for the key player on the changeover team. The champion is the person who runs the LCO team on a daily basis. This person could be a manager or, if appropriately supported, an operator or technician from the plant floor. The number one qualification is that they be passionate about changeover. Some might even say they need to be a bit nuts about it.

The champion must also be a recognized leader. This person will be responsible for keeping the team focused, following up to make sure that members complete their assigned tasks on time, requesting and justifying necessary resources, and reporting results to management. This seems like a job for a manager and it might be. In most plants there are also people at lower levels, such as mechanics, with these abilities. They should not be eliminated from consideration solely on the basis of position.

Virtually every department in a plant has a role to play on the changeover team. Some of these are obvious; others less so. The following identifies some other common departments or functions that need to be involved and why.

Changeover is carried out on the manufacturing floor, usually by the people working on the floor. This will be primarily mechanics, technicians, operators, and inspectors, though some plants will have others involved as well. As with the champion, passion, or at least a positive attitude, is probably the most important qualification. Experience and knowledge of the equipment, operating and regulatory requirements, as well as current practices are very useful. If available, it can be useful to seed the team with some members who have experience in other plants or industries. This provides more diversity of ideas.

Quality may need to perform pre- and post-changeover inspections to verify settings or cleanliness. These inspections can take time and there may be some ways to make them easier. If they need to inspect inside a machine cabinet for cleanliness, adding a light in the cabinet may eliminate the lost time they need to spend looking for a flashlight. In some plants there can be difficulty in getting inspections performed in a timely manner due to scheduling. A quality representative on the team will gain a better understanding of the issues that their department can cause and can suggest ways to improve them.

In most plants quality has a final veto over much of the operation. It is better to secure their ideas and cooperation up front for improvements. Seeking their approval after the fact can sometimes lead to miscommunication and other difficulties.

Engineering and maintenance have two LCO responsibilities. First, they must be aware of any operational or maintenance problems and take steps to correct them as well as prevent them from recurring. Second, it is likely that minor and possibly major modifications to machines or facilities will be required to reduce changeover time. Engineering will need to provide technical assistance to assure that these modifications are feasible and well-advised. They will need to update manuals and drawings to reflect the new conditions. Finally, they may need to provide the resources to carry out the modifications.

Purchasing and/or materials departments are responsible for obtaining the proper raw materials and components. Changeover improvement opportunities may arise based on how these come to the production floor—for example, individual cases versus bulk. Other opportunities may arise from purchasing higher-quality (less-variable) materials.

Marketing and product design determine what is to be made and its specifications. Part of the reason changeover exists is because of the explosion of product variety in the past 50 years. A product that might originally have been available in a single size and style may now be available in 50 variations. Each of these variations requires a changeover. There may be cases where the variation is insignificant from a marketing viewpoint, such as the glass vials with 0.63-inch difference in diameter discussed in Chapter 1. In other instances it may be possible to design bottles that look very different but minimize changeover by having identical touch points at top and bottom. Some packaging lines may be able to switch between cartons with either reverse or straight tucks. It is better if all cartons can be standardized to one or the other to eliminate the need to switch. The changeover team must have access to someone from the design department to discuss these issues.

Planning and scheduling may be done using an algorithm based on sales forecasts, warehouse levels, run-out time or other parameters. Companies that have families of similar products can sometimes reduce changeover events via scheduling. Scheduling similar products together can reduce or sometimes eliminate changeovers between them. Existing plant scheduling techniques may make this difficult to do. Planning needs to be part of the changeover team so that better scheduling techniques can be developed that meet customer needs while reducing changeover.

Finance must participate in the changeover project. Their primary function must be to develop a cost model that can be easily understood and used. Once developed, they must provide assistance to assure that the costs are kept up to date and that project justifications are properly done. Chapter 2 discussed the cost of changeover. The criticality of knowing changeover costs cannot be repeated often enough.

Human Resources (HR) has an important, if intermittent, role to play. A common goal is to involve operators more deeply in performance of changeover tasks. This change may require modifications to job descriptions. A plant that has a policy that does not permit operators to use tools may wish to reevaluate that policy. Training of operators, mechanics, and others may be required and this may be under the purview of HR. In some cases it might be desirable to change work schedules, such as moving cleaning to the third shift. This will require HR involvement.

If HR is responsible for company newsletters, bulletin boards, and other communication and recognition, they will need to get input from the changeover team for this.

Not all of these functions will exist in all plants. Some plants will have functions that were not mentioned. The previous discussion was not meant to be exhaustive. It was meant to demonstrate that virtually every department in a company has a role to play when it comes to reducing changeover time.

Any and all changes must be reviewed and approved by the safety department prior to implementation. It is better to discuss ideas for changes with the safety department before they get too far advanced. This helps avoid problems with approval in the future. LCO improvements must be approved no matter how slight and insignificant the change may appear to be. As important as changeover is, safety is even more important. It must always take first place above anything else.

Starting the Team

Once the team has been chosen, they need to be organized. A written charter from management is a useful tool. This can be developed before choosing the team, but another alternative is to begin with a loose, informal charter, then let the team develop a more formal charter. Allowing the team to develop their own charter from within helps cement them together and assure better buy-in. The charter should describe the purpose of the program and team, scope of work, how much time they are supposed to spend on it each week, and other details. A specific goal and time frame—for example, 50% changeover time reduction in 6 months—may be included in the charter. If it is not, the goal must be set out in writing in another document.

If possible, the team should be provided with their own meeting space. This will allow them to post drawings, timelines, and other visuals that they can leave up from one week to the next. They must also be provided with a specific time for meeting. This must be a reasonable block of time to allow the creative juices to get flowing. A 2-hour meeting every other week is probably better than a 1-hour meeting every week. Team members, and particularly their supervisors and managers, must understand that these meetings are a priority and that members must be permitted time from their regular duties to attend. Interruptions during the meetings must be kept to a minimum. The 3-minute phone call does not just take 3 minutes of the recipient's time. It derails the entire meeting and it may take 10–15 minutes to get it back on track.

Meetings must have a specific agenda. This should generally include status of current improvement projects, whether there is anything to report or not. Projects that are behind schedule must be evaluated to determine why and what actions are needed and by whom to get them back on schedule. As projects are completed, new projects can be discussed and assigned.

One of the first tasks of the team will be to identify changeover projects to pursue. This may sound easier than it actually is. One problem is that changeover, with everything that it affects and that is affected by it, can be quite overwhelming. This may make it hard to see where to begin. One good place is to develop SOPs and checklists of an existing changeover if they do not already exist. If the SOPs exist, they can begin with an audit to see how well they are being used. This eases them into the project, gives them a thorough understanding of existing practices, and gets them thinking about changeover in a systematic manner.

It is important to start slow. If unrealistic expectations are set on the first day, the team may become frustrated, causing loss of enthusiasm and effectiveness. Look for some easy quick hits. There are usually some improvements that can be implemented and completed within 10 working days. These can be a good place to start. These simple improvements may not cause dramatic results, but it gets the team used to success. It also gives the team something positive to show to the rest of the plant. Starting with a big project runs the risk of failure. Just as success tends to beget further success, failure tends to beget further failure. Begin with improvements that have low risk of failure and ramp up to more complex and significant (and riskier) projects.

The eventual goal is to reduce changeover time plantwide for all machines and processes. This may not be a reasonable initial goal. In addition to starting slow, start small. Identify a single machine, line, or process and concentrate initially on this. A narrow focus allows for maximum visibility of results. Not all attempts at improvement will be successful. Some ideas that looked good may, on implementation, turn out to have unforeseen consequences. Others might not result in as much savings as expected. It is best to find this out on a single line. As improvements demonstrate success, they can be copied to other lines.

There is also a human factor. Resistance to change is a natural human characteristic. Focusing on a single line reduces the number of people who need to be convinced. It is also easier to move people around if necessary to help assure positive attitude on the line. Others in the plant will see what is happening and the positive benefits. As they do, they will become more willing to buy in on their lines. The best circumstance is that they catch the enthusiasm and ask for the opportunity to improve their lines.

Publicize the results of the improvements as they occur; don't gunnysack them until the end of the year or even quarter. One possibility is to have a bulletin board in a visible area of the plant. One company has this board in the corridor leading to the cafeteria so that everyone sees it, or at least passes by it, daily. The board should have easy-to-understand pictures showing what was done, and graphs showing changeover times with trend lines. Congratulations should be given to team members who suggest ideas and other relevant information. If there is a company newsletter, the team should submit articles or pictures for it. Do use team members' names and pictures (unless they prefer not to have their names used, of course). Most people like to see themselves in print, especially in conjunction with a successful project benefitting the entire company.

The publicity should also include annual dollar savings wherever possible. Some management groups may view dollar savings as proprietary and be hesitant to share it on the plant floor. If it proves impossible to change this view, show time savings in the publicity. Include annual savings. Many people, even those directly involved, and even management, will not recognize the significance of saving 10 minutes on an improvement. They see this as about the amount of time they spend on a cup of coffee and how significant can that be? Ten daily minutes add up to more than 40 hours over the course of a year. Putting it in those terms makes the benefit much easier to see.

Who Performs Changeovers?

One issue that the changeover team will need to focus on eventually is the question of who should perform the changeovers. There are four main options:

- Operators
- Changeover specialists
- General mechanics
- Outside contractors

Many plants will have some combination of the first three or even all four. In general, operators should perform as many changeover tasks as possible for several reasons. First, they are available. Generally there will be considerably more operators in a plant than there are mechanics and they will generally cost less per hour. When the line is not running, they may have little else to keep them busy. It makes sense to use them on changeover tasks that they are capable of handling.

One objection to using operators for changeover is that they may not currently have the requisite skill levels. This is often true. That does not preclude them from being trained in certain skills. The goal is not to have them become mechanics but to give them the ability to perform the relatively simple, repetitive tasks of the typical changeover. As they are becoming more skilled, the changeover team will also be working to make the changeover simpler, better documented, and more repetitive. The combination of increased skills and reduced skill requirements will eventually meet in the middle.

Using operators to perform changeover makes them better operators. As they perform more tasks on the machinery, they will become more knowledgeable of how and why it works. This knowledge will allow them make minor adjustments and perhaps even minor repairs during operation, eliminating or reducing the downtime spent waiting for a mechanic.

One objection to using operators may come from the mechanics. They may feel that they are being replaced by the operators. In most plants there is far more work for mechanics in performing minor and major repairs, doing routine maintenance, and making improvements than there is time for them to do it all. Using the operators to perform the routine work of changeover frees the mechanics for more valuable tasks. It allows them to use their unique skills in more highly value-adding ways instead of wasting them on routine changeover. It is important to explain this to them so they do not feel like they are being set up to be made redundant.

A second alternative for changeover is to use setup mechanics. These go by a variety of names, but they are basically a lower, or even entry level mechanic. In addition to changeovers, they can also perform preventive maintenance and some of the simpler repairs. The position should generally serve as training and a step on the ladder to a more highly skilled general mechanic.

General mechanics are those who by education, training, and/or experience are qualified to perform most adjustments, maintenance, and repairs to the machinery. Their skills may be needed at present when changeover requires a lot of judgment and skill. As changeover is simplified and made more repetitive, they become overqualified for these tasks and their skills go to waste. Freed from changeover, their focus should shift to other, higher-order tasks, including troubleshooting, repair, and improvement.

One higher-order task will be improving changeover to the point where operators and others can do it. These mechanics usually have an enormous amount of valuable knowledge. It is important to get them working with the changeover team as members or at least cooperators to take advantage of this knowledge.

A final possibility is to use an outside contractor to perform changeover. This may be a useful option in several scenarios:

■ Changeover is infrequent and requires specialized skills and/or equipment. Periodic chemical cleaning of food processing equipment is one example.
■ Changeover is frequent but requires fairly generic skills. Contract custodial services for cleaning of offices and other general areas are very common. Some companies go a step further and use these contractors for cleaning production areas during changeover.
■ Some companies use a contractor to supply mechanics to carry out routine changeovers that would normally be done by in-plant personnel.

There are some benefits to using outside contractors, but there can be drawbacks as well. The principle one is *control*. A company never has as high a level of control over the selection, training, and direction of contract employees as it will over their own. This may not be a big issue for routine work such as cleaning, but it may be necessary in the case of specialized services.

A related issue is scheduling. The contractor may have other commitments that may not permit the flexibility that comes from using a plant's own employees. As plant priorities shift, it may be necessary to pull people off of one set of tasks to put them on another set. Shifting people around like this may be harder when they belong to the contractor rather than being under direct plant control.

Another consideration, regardless of which of these options is selected, is organizing the changeover in the plant. One possibility is to have a changeover team. This team will consist of a group of mechanics and operators and perhaps others with the skills needed for changeover. They will be shifted from line to line as needed, focusing always on changeover. The advantage is that by focusing on changeover, they will gain a depth and breadth of knowledge that should make them more effective at it. The disadvantage is that this experience comes at the expense of those normally working on the line who might otherwise perform the changeover.

Another potential drawback can be scheduling. When the changeover team is changing over one line, another line may be sitting idle waiting for them to become available.

Getting to Work

After the team and the focus have been selected, it is time to get to work. The first step in improving changeover is an understanding of the current process. This understanding is best gained by observing a routine change-over. If there are several types of changeover, such as minor and major, the team must decide which to observe. During the observation, they must be observers, not participants. If they are doing the changeover, it is hard for them to see what was done. As observers, they should try to interrupt the teammates performing the work. The mere fact of observing a task can change the way in which it is carried out. LCO team members may need to ask questions or interrupt a task to take pictures. This is all necessary, but the goal is to cause as little disruption as possible.

Prior to beginning the observation, it is important to do some preparatory work with the teammates performing the changeover. It can be unnerving to have a team descend on one to watch their work. This is especially true when there are stopwatches, cameras, and videos involved. If there is an existing SOP and they do not normally follow it exactly, they may worry that they will get in trouble for doing a normal changeover. Ideally, they should be aware of the project and what it is planned to accomplish. In all cases, they need to be assured that they have near total immunity while performing their job as they normally would. They need to be assured that the observations are to collect data on the process and not to determine how well they perform their jobs. As the ones who normally perform the changeover, they probably have had some ideas on how they could be performed better. They need to be encouraged to share these ideas with the team. The team should incorporate these ideas, even if they may prove unfeasible for whatever reason. If they are feasible, especially simpler ideas, they should be implemented as quickly as possible. This will help convince the teammate that their opinion is important.

Some teammates will appreciate credit for the suggestions, and if so, they should be mentioned by name, for example, "Jenny Jones suggested that..." Others may prefer anonymity and the idea should be unattributed, for example, "It was suggested that..." It is important that the LCO team members know what the individual's preference is. If unknown, it is probably better to err on the side of anonymity.

Everyone should be provided with a notepad to write down any observations or ideas. These may be a bit informal and a team member needs to be designated as the official note taker. They will make a detailed list of all tasks performed in the changeover. *Detailed* means that each task will be broken down into subtasks that are as basic as possible.

They should not write "Remove the starwheels" except as a general heading. They will instead write:

- Remove 3 bolts holding the starwheel to the hub.
- Set the bolts in the provided tray.
- Move the guide rail out of the way.
- And so on.

The reason for this detailed breakdown is that it may be hard to see how to improve removing the starwheel. It may be much easier to come up with ideas to simplify or eliminate the bolt removal or moving the guide rail.

If more than one set of changeover tasks is being performed at a time, it may be necessary to have two recorders, one on each set.

Another member of the team should be identified as the timekeeper and provided with a clipboard with two stopwatches on it. One stopwatch will record total changeover time. The other will be used to time tasks.

Another LCO team member should have a camera to take pictures. Digital pictures are cheap and it is important to take as many as possible. Pictures of no value can always be discarded. Use a zoom lens to get in close to where the work is being done as well as framing shots to show overall context. When taking pictures, be careful not to take any pictures of teammates without their permission. If necessary, ask them to step back from the task while the picture is taken.

Another member should have a floor plan and be making a "spaghetti diagram" showing the movements of the people doing the changeover.

This sounds like a lot going on all at once and it certainly can be. It may be necessary to observe two or more changeovers to capture everything. A video camera is also very useful. As with the camera, it is important not to get anyone in the video without their permission. Usually people will not have a problem with it if asked. The problem comes when they turn up in a video for which they have not given permission.

When shooting the video, it is not necessary to make it a Hollywood production. In fact, too much video activity can be more of a distraction than a help. Mount the video on a tripod, aim it at the area of interest, zoom it in as appropriate, and let it run. If necessary it can be removed from the tripod for better video of a task, but in general it should be left alone. If all work stops while the mechanic fetches a part, leave the video on. This will provide a record of how much time was lost chasing parts and provide more impetus for externalization. If the camera has a clock that can be shown on the screen, activate this to provide a detailed and accurate timing record.

Other videos may be shot in the future for specific purposes, such as training, and more care may be necessary. The purpose of this video is to provide a record of what was done. If something is not clear on the video, it is likely that someone will recognize the task and be able to fill in any necessary details.

After completion of the changeover, it can be mapped. There are several project management programs that can be used to make Program Evaluation and Review Technique/Critical Path Method (PERT/CPM) charts of the changeover process. An alternate, perhaps better, technique is to get a roll of brown butcher paper and hang a long sheet on the wall. Sticky notes are used to identify all the tasks in sequence. Tasks that are done in parallel are identified on additional timelines on the same sheet. Tasks can be written directly on the paper, but the sticky notes allow more flexibility since they can be easily moved or changed. Color-coded notes can also be used to track different tasks or different teammates performing the changeover. Specific times can be identified for activities or the overall map can be sectioned into time buckets. The length of the buckets will depend on overall changeover time as well as how finely it is desired to divide the timings. At this point in the LCO project, dividing the time too finely may prove to be more of a distraction than a benefit. Fifteen- or 30-minute buckets may be appropriate for most changeovers taking hours. Changeovers that take less than an hour might be best divided into 5- or even 1-minute buckets.

Now that the actual, as opposed to the ideal, changeover has been documented and mapped, the team can begin looking for improvements. Some of these will be immediately apparent quick hits that can be implemented in days not weeks. As noted previously, go ahead and get these knocked out if for no other reason than to create a winning feeling.

Then apply the ESEE technique discussed in previous chapters: Look first for tasks that can be eliminated. Question everything. Ask "Why is XXX being done?" If there is not a good reason, stop doing it. Resources that are spent improving an unnecessary task are wasted resources that could be better used elsewhere.

Simplify everything as much as it possibly can be. This book has discussed a number of simplification techniques. The reader will undoubtedly think of others. Identify where they can be used and implement them.

Externalize, externalize, externalize. The only reason the plant exists is to produce. When the line is down for changeover, it is not producing. Specifically, it is not making money. Many tasks can be done while the line is running. It may take a little doing, but it will almost always pay off. Shift everything possible from internal to external.

And finally, execute. Execute the changeover exactly the same way every time. The closer one can come to zero variation in changeover, the closer one can come to eliminating the wastes of startup time. The better the execution, the better the line will start up and the better it will run.

As each opportunity is identified, develop an action plan for what is to be done and how. Unless it also has an assigned responsibility and due date, it is not an action plan, so be sure to assign these as well. Finally, do a cost–benefit analysis. Identify how much the improvement will cost in out-of-pocket expense and how much downtime it will eliminate. If there is a dollar cost per hour for changeover, use that to calculate payback on the expenditure as well as total annual savings.

Conclusion

Just as lean manufacturing must be a way of life for plants to be successful and thrive, so must lean changeover. In stock car racing they say "If you ain't running, you ain't racing," meaning don't spend any more time in the pits than absolutely necessary. The same concept applies to changeover. "If you ain't running, you ain't making any money."

Treat changeovers like pit stops. Even small incremental improvements will make you a winner.

Appendix

STANDARD OPERATING PROCEDURE			
Title:		Page 1 of 3	
Written by:		SOP#	
Approved by:		Rev. #:	
Approval date:			
Effective date:			

This form design may be used for writing standard operating procedures (SOPs) for various types of equipment. It is a generic form and may be modified to fit the particular needs of any organization or type of SOP. Once a specific SOP format has been developed, it is suggested that all sections be used in all SOPs. If a section is not applicable, a simple N/A may be inserted. This assures that nothing is missed. Annotations in italics describe the purpose of each section.

1. PURPOSE
    This section is a simple description of 1–3 sentences of the purpose of the SOP.

2. GENERAL INFORMATION
    2.1 Scope
        This section is used to describe precisely which equipment is covered by the SOP and under what circumstances.
    2.2 Safety
        All applicable safety steps and precautions should be addressed here. Examples could include use of safety glasses and other

STANDARD OPERATING PROCEDURE		
Title:		Page 2 of 3
Written by:	SOP#	
Approved by:	Rev. #:	
Approval date:		
Effective date:		

> *safety equipment or lockout/tagout procedures. Where there are separate safety SOPs, they should be referenced by number here.*

2.3 Responsibility

> *This section describes who is responsible, by job skill, department, or other classification, for performing the SOP.*

3. MATERIALS

This section lists all changeparts required for the changeover. If components or product are necessary to changeover, they should be noted here.

4. TOOLS

List all tools that will be required for the changeover.

5. DEFINITIONS

Any nonstandard terms or expressions should be defined in this section.

6. PROCEDURE

This is the heart of the SOP. In this section the entire process of disassembly and reassembly will be covered step by step. The goal should be to describe it fully enough that anyone with a minimal amount of machine knowledge can carry out the changeover. The actual procedure may use as many steps and substeps as may be required for clarity. Pictures should also be incorporated.

6.1

6.1.1

6.1.1.1

7. DOCUMENTATION

This section describes the various forms and other documentation to be used with the SOP.

8. ATTACHMENTS

This section is for any additional information that might be helpful.

STANDARD OPERATING PROCEDURE			
Title:		Page 3 of 3	
Written by:		SOP#	
Approved by:		Rev. #:	
Approval date:			
Effective date:			

8.1 Machine settings

This section is a tabular listing of all machine settings for each product or package size. It will include not only various machine components but any temperature or pressure settings as well.

8.2 Machine layout drawings

This section will show a conceptual layout of the machine with all adjustment points clearly marked. The location of all change-parts on the machine should be clearly identified.

9. REFERENCES

This section can be used to list other SOPs, equipment manuals, drawings, or any other information that may be useful.

10. CHECKLIST

Changeover SOPs should be complete and thorough. In order to enhance usability, a checklist should be included as a final attachment. This checklist should always be used by the person doing the changeover to assure that all steps have been carried out in proper order. One way to automatically develop a checklist is to use major MS-Word outline headings for action items with subheadings for describing them in detail. A secondary table of comments, on a separate page, can then be generated to show the headings as well as the page on which the complete description appears. For example:

6.1 Set the label sensitivity

6.1.1 Move label and backing in front of the photoeye.

6.1.2 Adjust the trim pot until the light is on.

6.1.3 Note trimpot position.

6.1.4 Etc.

Section 6.1, standing alone, becomes a checklist item.

6.1 Set the label sensitivity 4

For more information visit www.changeover.com
Contact info@changeover.com or 787-550-9650

Changeover.com offers workshops, assessments,
and other services to assist in reducing changeover time

Changeover annual loss calculator

Less Red, More Green!

Calculate your savings from a lean changeover (LCO) program by Changeover.com

COMPANY DATA

HOW MUCH CAN LCO SAVE YOU?

A.	Changeovers per week (#)	5	Enter your changeover reduction target (%) below
B.	Average changeover time including startup (minutes)	120	
C.	Production weeks per year (#)	50	
D.	Shift length (hours)	8	25%
E.	Normal production hours per week (hours)	40	
F.	Average production rate per minute (PPM)	500	
G.	Average value per product ($)	$0.50	
H.	Cost of changeover downtime ($/hr)	$10,000.00	

TIME LOSS

TIME GAIN

I.	Annual lost production from changeover (minutes)	(30,000)	7,500.0 Minutes
J.	Annual lost production from changeover (hours)	(500.0)	125.0 Hours
K.	Annual lost production from changeover (shifts)	(62.5)	15.6 Shifts
L.	Annual lost production from changeover (weeks)	(12.5)	3.1 Weeks

PRODUCTION LOSS

PRODUCTION GAIN

M.	Annual non-produced products (quantity)	(15,000,000)	3,750,000 Products
N.	Annual non-produced products ($ value)	($7,500,000.00)	$1,875,000.00 Dollars

DOLLAR LOSS

DOLLAR GAIN

O.	Annual cost of changeover ($)	($5,000,000.00)	$1,250,000.00 Dollars

To use this calculator, replace the numbers shown for illustration in the "Company Data" section with actual or estimated values. Calculations of time, production and dollar losses will calculate automatically. Other currencies may be entered. Results will be calculated in those currencies though they will show a dollar sign.

Typically it is possible to reduce changeover time from 25 to 75% in 6 months with no major capital expenditure upon implementation of an effective lean changeover (LCO) reduction program by Changeover.com

Call or e-mail John Henry, the Changeover Wizard, at 787-550-9650 johnhenry@changeover.com to find out how.

Pass it on! This calculator is ©2011 by John R Henry but may be freely copied and distributed provided that proper credit is given and no contact information is removed.

Index